BRITISH ASSOCIATION FOR THE ADVANCEMENT OF SCIENCE

MATHEMATICAL TABLES
PART-VOLUME B

# THE AIRY INTEGRAL

### GIVING TABLES OF SOLUTIONS OF THE DIFFERENTIAL EQUATION

$$y'' = xy$$

PREPARED BY

## J. C. P. MILLER

ON BEHALF OF THE COMMITTEE FOR THE
CALCULATION OF MATHEMATICAL TABLES

*Published for the British Association*

AT THE

UNIVERSITY PRESS
CAMBRIDGE
1946

# CAMBRIDGE
## UNIVERSITY PRESS

University Printing House, Cambridge CB2 8BS, United Kingdom

Cambridge University Press is part of the University of Cambridge.

It furthers the University's mission by disseminating knowledge in the pursuit of education, learning and research at the highest international levels of excellence.

www.cambridge.org
Information on this title: www.cambridge.org/9781316611951

© Cambridge University Press 1946

First published 1946
First paperback edition 2016

A catalogue record for this publication is available from the British Library

ISBN 978-1-316-61195-1 Paperback

# INTRODUCTION

## 1. *Historical*

The function with which these tables are mainly concerned seems to have been investigated first by Airy* (1838, 1849) and has been, in consequence, named after him. In calculating light-intensity in the neighbourhood of a caustic, Airy met with the integral

$$W = \int_0^\infty \cos \tfrac{1}{2}\pi (w^3 - mw)\, dw$$

He gave (1838, p. 390) a 5-decimal table of $W$ for $m = -4\cdot0(0\cdot2) + 4\cdot0$, and a table of $W^2$ to 5 or more decimals over the same range of $m$. These tables were derived from 7-decimal calculations of $W$, the values being given in full on page 402 of the same paper. Later (1849, p. 598) he gave a newly-calculated 5-decimal table of $W$ for $m = -5\cdot6(0\cdot2) + 5\cdot6$.

The 1838 table was calculated by quadratures supplemented by an expansion, asymptotic in character, for the 'tail' of the integral. For the 1849 table the ascending series was used. Airy's reasons for his choice of methods are not uninteresting (1849, p. 595):

"The computation by quadratures was exceedingly laborious, and I did not resort to it without trying other methods of a more refined nature. But in every attempt at expansion of the formula I was met by the integral of a sine or cosine with infinite limits. The reasonings upon which several mathematicians have attempted to establish the value of such an integral appeared to me so little conclusive, that I preferred at once to abandon the expansions which introduced them, and to rely only on the infallible but laborious method of quadratures.

"On my stating to Professor De Morgan, after terminating the calculations, the scruples which had led me to reject the expansions, he expressed himself so strongly confident of the correctness of the conclusions upon the point which I had considered doubtful, that I was induced to undertake the numerical computation of the series given by expansion of the formula."

That Airy's scruples were not entirely unjustified is apparent on examination of the various attempts to attach a meaning to $\sin\infty$ and $\cos\infty$, and to integrals of the type to which he refers.

Having performed the new calculations, he draws the conclusion (1849, p. 599):

"The agreement of the values of the integral, computed by methods so totally different, is not a little remarkable. On the one hand, it may be received by some persons as a proof of the correctness of that part of the theory of series which asserts the evanescence of the integral of a cosine when the limits are 0 and 1/0: on the other hand it may be considered to afford evidence of the great care with which the quadrature computations had been made."

A recalculation of Airy's values has been made, using the present tables, and noting that

$$W = 2\kappa\, \mathrm{Ai}(-\kappa m)$$

where $\kappa = (\pi^2/12)^{1/3}$. This constant $\kappa$ is characteristic of relations between Airy's form of his integral and the definition adopted in this work; the values of $\kappa$ and of its reciprocal are, to ten decimals,

$$\kappa = 0\cdot93692\ 78888 \qquad\qquad 1/\kappa = 1\cdot06731\ 79996$$

The recalculation has revealed that, in units of the seventh decimal, the greatest error in any of Airy's 7-decimal values for $-4\cdot0 \leqslant m \leqslant +2\cdot0$ is 52 units; the error then changes fairly steadily from $-56$ units at $m = 2\cdot2$ to $-272$ units at $m = 3\cdot4$; at $m = 3\cdot6$, $3\cdot8$ and $4\cdot0$ the errors are $-23$, $+858$ and $+2661$ of these units. Airy's revised table has only one error of more than 2 units of the fifth decimal; the value for $m = -3\cdot6$ is 3 units in error, the value in the original table being, however, correct to 5 decimals.

The ascending series, although an enormous improvement compared with quadratures, is still very laborious when $m$ is large, and the desire to reduce this labour induced Stokes (1851, 1858, 1907) to develop asymptotic expansions and led to his remarkable discovery of the discontinuity of the

---

* Dated references are given in the Bibliography.

'arbitrary constants' appearing in asymptotic developments (1858). In the 1851 paper Stokes also developed asymptotic expansions for zeros of $W$ and of its derivative, and tabulated, to 4 decimals, the first 50 zeros of $W$ and the first 10 zeros of its derivative. These have been recomputed using the present tables; they are respectively $-a_s/\kappa$ and $-a_s'/\kappa$, where $a_s$ and $a_s'$ are the zeros of Ai$(x)$ and Ai$'(x)$, see Table III, and $1/\kappa$ has the value already given. With one exception (the first zero of the derivative, which should be 1·0874 not 1·0845), all Stokes's values are correct within a unit of the fourth decimal.

Airy's and Stokes's tables have been reproduced several times (see Fletcher, Miller and Rosenhead, *Index of Mathematical Tables*, Sub-section 20·2, London, Scientific Computing Service), but, except for a small table of $\sqrt{\pi}$ Ai$(-x)$ in Kramers (1926, p. 840), which has an unreliable third decimal, no new calculation seems to have been made until Jeffreys (1928, p. 107) announced that he had made a table of Ai$(x)$ and Ai$'(x)$; this table is for $x = -2\cdot05(0\cdot05) + 2\cdot05$, with 8 working decimals, and has been incorporated in the present tables. There are, however, tables of the closely related Bessel functions of order $\pm 1/3$ and $\pm 2/3$ and of their zeros (*Index*, Articles 17·221–17·232, 17·752–17·7536, 18·221–18·222); the most extensive of these yet published is in Watson (1922), which gives (pp. 714–729) $J_{1/3}(x)$, $Y_{1/3}(x)$, $|J_{1/3}(x) + iY_{1/3}(x)|$, $e^x K_{1/3}(x)$ to 7 decimals, and $\tan^{-1}\{Y_{1/3}(x)/J_{1/3}(x)\}$ to $0''\cdot01$, all for $x = 0\cdot00(0\cdot02)16\cdot00$, and (p. 751) the first 40 zeros of $J_{1/3}(x)$, $Y_{1/3}(x)$, $J_{-1/3}(x) \pm J_{1/3}(x)$ to 7 decimals. Watson also notes that to compute functions of order $-1/3$ the phase should be increased by $60°$. A more extensive MS. table prepared by the Mathematical Tables Project of the New York W.P.A. has been announced (see *Mathematical Tables and other Aids to Computation*, **1**, 93, 1943); the main tables give $J_{\pm 1/3}(x)$, $J_{\pm 2/3}(x)$, $I_{1/3}(x)$ and $I_{2/3}(x)$ for $x = 0\cdot00(0\cdot01)25\cdot00$, and $I_{-1/3}(x)$ and $I_{-2/3}(x)$ for $x = 0\cdot00(0\cdot01)13\cdot00$, all to ten decimals or figures. The connections between these functions and the functions Ai$(x)$ and Bi$(x)$ are exhibited on pages B9 and B17. Another table of interest is a short (but apparently unique) table for pure imaginary argument given by Rayleigh (1915).

There have been several theoretical investigations of the properties of the Airy Integral and of allied functions; in particular those of Nicholson (1909), of Brillouin (1916) and of Kramers (1926) may be noted. A number of their results, and others, are given in Watson (1922, pp. 188–190, 248–252, 320–324).

Recently the demand for tables of the Airy Integral has revived; this revival is closely connected with the simplicity of the differential equation satisfied by the function. We readily verify that

$$\frac{d^2W}{dm^2} = -\frac{1}{12}\pi^2 m W = -\kappa^3 m W$$

Jeffreys (1928, p. 105, but with a later change in the sign of $x$) has introduced changes of scale in function and argument, defining*

$$\text{Ai}(x) = \frac{1}{\pi}\int_0^\infty \cos\left(\tfrac{1}{3}t^3 + xt\right) dt \tag{1}$$

so that, using accents to denote differentiation with respect to the argument, Ai$(x)$ satisfies the differential equation

$$y'' = xy \tag{2}$$

This differential equation arises naturally as an approximation to the general second order differential equation over a limited range of the argument, for, supposing such an equation to be reduced to the normal form (see, e.g., Ince, *Ordinary Differential Equations*, p. 394, 1927)

$$y'' + I(x)\,.\,y = 0$$

we may in general approximate to $I(x)$ in the neighbourhood of $x = x_0$ by an expression of the form

$$I(x) = a + b(x - x_0) = I(x_0) + (x - x_0)\,I'(x_0)$$

neglecting terms involving higher powers of $x - x_0$. A change of origin and scale for $x$ now leads to the equation (2). This method of approximation is especially useful near a zero of $I(x)$, i.e. when $a = 0$; see Jeffreys (1942). It soon became apparent that tables of a suitable second and independent solution of the differential equation were also needed.

* Watson, in his discussions, deals with the functions $\int_0^\infty \cos(t^3 \pm xt)\,dt = 3^{-1/3}\pi\,\text{Ai}(\pm 3^{-1/3}x)$ which satisfy the differential equations $d^2y/dx^2 = \pm\tfrac{1}{3}xy$.

# CONTENTS

## 2.  *Description of the Tables*

2·1.  The complete integral of the differential equation (2) may be written in the form

$$y = A \, \text{Ai}(x) + B \, \text{Bi}(x) \tag{3}$$

where $A$ and $B$ are constants, and Bi($x$) is a suitably chosen independent solution of (2) that is defined in §4. The tables are concerned mainly with Ai($x$) and its derivative; every solution of (2) which remains finite as $x \to \infty$ is a multiple of this integral, which itself tends to zero as $x \to \infty$.

The complete integral of (2) can also be written in the form

$$y = C \, F(x) \sin \{\chi(x) + \epsilon\} \tag{4}$$

in which $C$ and $\epsilon$ are arbitrary constants. If we take

$$C^2 = A^2 + B^2 \qquad\qquad \tan \epsilon = B/A \tag{5}$$

where $A$ and $B$ are the same constants as in (3), then

$$\text{Ai}(x) = F(x) \sin \chi(x) \qquad\qquad \text{Bi}(x) = F(x) \cos \chi(x) \tag{6}$$

The corresponding derivative can be similarly expressed, with the *same* constants $C$ and $\epsilon$, as

$$y' = C \, G(x) \sin \{\psi(x) + \epsilon\} \tag{7}$$

where $\qquad\qquad \text{Ai}'(x) = G(x) \sin \psi(x) \qquad\qquad \text{Bi}'(x) = G(x) \cos \psi(x) \tag{8}$

The tables aim at an 8-figure standard of accuracy throughout for Ai($x$), Bi($x$) and their derivatives; this corresponds to 6 decimals of a degree in the phases $\chi(x)$ and $\psi(x)$ in Table VII, and to 9 decimals for $\pm \tau = 0 \cdot 1$ Bi$'(\pm x)$ in Table IV. Reduced derivatives of higher order in Table IV are given to the 10 decimals needed for interpolation of Bi$'(x)$ to 8 decimals.

Provision for interpolation is made everywhere (except of course in Tables III and V); the desire to make this provision resulted in a decision to tabulate log$_{10}$ Ai($x$), Ai$'(x)$/Ai($x$), log$_{10}$ Bi($x$) and Bi$'(x)$/Bi($x$) in Tables II and VI. The details of interpolation are discussed in § 3; BRITISH ASSOCIATION *Auxiliary Table I* is available to assist in this process.

2·2.  *Notation for Reduced Derivatives.* The operator $\tau$ is defined by

$$\tau \phi(x, h) = \int_0^h \frac{\partial}{\partial x} \phi(x, t) \, dt \tag{9}$$

If this operator is applied to the function $f(x)$, which is the same for all values of $h$, it is readily verified by repeated application that

$$\tau^n f(x) = (h^n/n!)\, f^{(n)}(x) \tag{10}$$

that is, $\tau^n f(x)$ is the $n$th reduced derivative of $f(x)$, a term in the Taylor expansion of $f(x + h)$. For this reason $\tau$ may be called the *Taylor Operator*. It should also be noted that

$$(-\tau)^n f(-x) = (h^n/n!)\, f^{(n)}(-x) \tag{11}$$

In these tables $\tau^n f(x)$ is denoted by $\tau^n$ when dealing with a specific function such as $\mathrm{Bi}(x)$ or $\mathrm{Bi}(-x)$, as in the headings of Table IV and in formulae (23) to (26). In order to emphasize the factor $(-1)^n$ in (11) and to enable the reduced derivatives to be given throughout with the correct sign (i.e. that of the derivative) $(-\tau)^n$ is given in Table IV when the argument is negative, that is, for $\mathrm{Bi}(-x)$ with $x > 0$. Since $h = 0.1$ in this table, the columns headed $\tau$ and $-\tau$ thus give always $+0.1\,\mathrm{Bi}'(\pm x)$.

2·3. *The Phases* $\chi(x)$ *and* $\psi(x)$. These angles are tabulated in degrees, to simplify determination of their sines and cosines. Decimals of a degree are given, and multiples of $360°$ have been subtracted from the angles; this multiple, although not usually needed, is also given.

The following tables give natural values of sines and cosines with argument in degrees and decimals; all include convenient provision for interpolation.

BUCKINGHAM, E. *Manual of Gear Design*. Section One, *Mathematical Tables*. New York, Machinery Publishing Co., 1935. This gives 8-decimal values for arguments $0°\!·\!00(0°\!·\!01)45°\!·\!00$; see Comrie, *Mathematical Tables and other Aids to Computation*, **1**, 88, 1943 on errors (few and small in the tables of sines and cosines).

HERGET, P. *Astron. Journal*, **42**, 123–125, 196, 1933. This is a one-page table of 8-decimal values for arguments $0°(1°)45°$.

PETERS, J. *Siebenstellige Werte der trigonometrischen Funktionen*... Berlin-Friedenau, Optische Anstalt C.P. Goerz, 1918. (Since 1930, Leipzig, Teubner; reprint, New York, Van Nostrand, 1942.) This gives 7-decimal values of $\sin\theta$ for arguments $0°\!·\!000(0°\!·\!001)90°\!·\!000$.

## 3. *Interpolation*

If full use is made of all differences, modified differences or reduced derivatives printed, the resulting interpolated value will be correct within $1\tfrac{1}{2}$ units of the last figure tabulated. Again, if this accuracy is desired, then, in general, all differences or modified differences that are given must be used; on the other hand, as many reduced derivatives are tabulated as are needed to obtain interpolated values of $\mathrm{Bi}'(x)$ from either end of the interval, and these will not all be needed if only the nearer tabular point is used, or if a value of $\mathrm{Bi}(x)$ is sought. Details of the interpolation methods suggested are set out below.

3·1. *Interpolation by Differences.* In Tables I, II, VI and VII second differences are given for interpolation with Everett's formula,

$$f_\theta = f(x + \theta h) = \phi f_0 + \theta f_1 + E_0^2 \delta_0^2 + E_1^2 \delta_1^2 \tag{12}$$

where $h$ is the tabular interval, $\phi = 1 - \theta$, and terms involving fourth and higher differences are omitted.

In some cases these omitted differences are not negligible, and the use of (12) may entail substantial end-figure error. In such cases slight modifications of Everett's formula are advocated, and means for their use provided.

If $\delta^4$ is not greater than 1000 it may be allowed for by the "throw-back*", using the formula

$$f_\theta = \phi f_0 + \theta f_1 + E_0^2 \delta_{m0}^2 + E_1^2 \delta_{m1}^2 \tag{13}$$

where $\delta_{m0}^2$ and $\delta_{m1}^2$ are the "modified second differences" of $f_0$ and $f_1$ respectively (see (16) below); (13) should be used where these, *and no other differences*, are provided. The greatest possible error, excluding rounding-off errors made during the interpolation, is less than a unit of the last figure tabulated.

---

* Comrie, *Interpolation and Allied Tables*, p. 928, 1936, London, H.M. Stationery Office. An elaboration of this idea is used below. See also BRITISH ASSOCIATION, *Mathematical Tables*, Vol. I, p. xi, 1931.

There still remain short ranges of the argument where fourth and higher differences are of such magnitude that values given by (13) may be several units out in the end figure. The error can everywhere be reduced to less than $1\frac{1}{2}$ of these units by one or other of the formulae

$$f_\theta = \phi f_0 + \theta f_1 + E_0^2 \delta_{m0}^2 + E_1^2 \delta_{m1}^2 + M_0^4 \gamma_0^4 + M_1^4 \gamma_1^4 \tag{14}$$

$$f_\theta = \phi f_0 + \theta f_1 + E_0^2 \delta_{m0}^2 + E_1^2 \delta_{m1}^2 + T^4(\gamma_0^4 + \gamma_1^4) \tag{15}$$

where (the coefficient 0·184 being exact)

$$\left.\begin{aligned}
\delta_m^2 &= \delta^2 - 0\cdot184\delta^4 + 0\cdot038082\delta^6 - 0\cdot00830\delta^8 + 0\cdot0019\delta^{10} - 0\cdot0004\delta^{12} + \ldots \\
&= h^2 f'' - 0\cdot100667 h^4 f^{iv} + 0\cdot010193 h^6 f^{vi} - 0\cdot00103 h^8 f^{viii} + 0\cdot00010 h^{10} f^x - \ldots \\
1000\gamma^4 &= \delta^4 - 0\cdot27827\delta^6 + 0\cdot0685\delta^8 - 0\cdot0164\delta^{10} + 0\cdot004\delta^{12} - \ldots \\
&= h^4 f^{iv} - 0\cdot111603 h^6 f^{vi} + 0\cdot01142 h^8 f^{viii} - 0\cdot0012 h^{10} f^x + \ldots
\end{aligned}\right\} \tag{16}$$

and

$$M^4 = 1000(E^4 + 0\cdot184 E^2) \qquad 2T^4 = M_0^4 + M_1^4 \tag{17}$$

The necessity for the use of (14) or (15) is indicated by the tabulation of $\gamma^4$; if $|\gamma_1^4 - \gamma_0^4|$ averages more than a unit (14) must be used, otherwise (15) may be simpler and is subject (in such cases only) to a slightly smaller maximum error.

The coefficients $E_0^2$, $E_1^2$, $M_0^4$, $M_1^4$ and $T^4$ are given separately as BRITISH ASSOCIATION *Auxiliary Table I*, for arguments $\theta = 0\cdot00(0\cdot01)1\cdot00$. The following table of $M^4$ and $T^4$ may, however, be found helpful.

| $\theta$ | $M_0^4$ | $M_1^4$ | $T^4$ | $\theta$ | $M_0^4$ | $M_1^4$ | $T^4$ |
|---|---|---|---|---|---|---|---|
| 0·0 | −0·000 | +0·000 | −0·000 | 0·5 | +0·219 | +0·219 | +0·219 |
| 0·1 | −0·698 | +0·256 | −0·221 | 0·6 | +0·448 | −0·128 | +0·160 |
| 0·2 | −0·768 | +0·448 | −0·160 | 0·7 | +0·523 | −0·506 | +0·009 |
| 0·3 | −0·506 | +0·523 | +0·009 | 0·8 | +0·448 | −0·768 | −0·160 |
| 0·4 | −0·128 | +0·448 | +0·160 | 0·9 | +0·256 | −0·698 | −0·221 |
| 0·5 | +0·219 | +0·219 | +0·219 | 1·0 | +0·000 | −0·000 | −0·000 |

3·2. *Interpolation by Reduced Derivatives.* In terms of the operator $\tau$, defined in (9), Taylor's expansion for $f(x + \theta h)$ may be written

$$f(x + \theta h) = (1 + \theta\tau + \theta^2\tau^2 + \ldots + \theta^n\tau^n + \ldots)f(x) \tag{18}$$

It is of interest to note, since

$$h\,\partial f(x + \theta h)/\partial x = \partial f(x + \theta h)/\partial\theta \tag{19}$$

that

$$\theta\tau f(x + \theta h) = f(x + \theta h) - f(x) \tag{20}$$

whence

$$(1 - \theta\tau)f(x + \theta h) = f(x) \tag{20}$$

in formal agreement with the expansion (18).

Differentiation and integration of (18) with respect to $\theta$, with use of (19), give Taylor expansions for the derivative and integral of $f(x)$ at an arbitrary point:

$$hf'(x + \theta h) = (\tau + 2\theta\tau^2 + 3\theta^2\tau^3 + \ldots + n\theta^{n-1}\tau^n + \ldots)f(x) \tag{21}$$

$$\int_x^{x+\theta h} f(t)\,dt = h\left(\theta + \tfrac{1}{2}\theta^2\tau + \tfrac{1}{3}\theta^3\tau^2 + \ldots + \frac{1}{n}\theta^n\tau^{n-1} + \ldots\right)f(x) \tag{22}$$

The formulae (18) and (21) are to be used for interpolation in Table IV. Expressed in terms of the shortened notation mentioned in §2·2 these become

$$f(x + \theta h) = f(x) + \theta\tau + \theta^2\tau^2 + \ldots + \theta^n\tau^n + \ldots \tag{23}$$

$$hf'(x + \theta h) = \tau + 2\theta\tau^2 + 3\theta^2\tau^3 + \ldots + n\theta^{n-1}\tau^n + \ldots \tag{24}$$

Similarly (cf. (11))

$$f\{-(x + \theta h)\} = f(-x) + \theta\tau + \theta^2\tau^2 + \ldots + \theta^n\tau^n + \ldots \tag{25}$$

$$-hf'\{-(x + \theta h)\} = \tau + 2\theta\tau^2 + 3\theta^2\tau^3 + \ldots + n\theta^{n-1}\tau^n + \ldots \tag{26}$$

where $f'(-x) = \{df(t)/dt\}_{t=-x} = -df(-x)/dx$. Again it must be noted that in Table IV the reduced derivatives are given with the true sign of the derivative, so that, when the argument $-x$ is negative, the quantity $(-\tau)^n = (h^n/n!)\{Bi^{(n)}(t)\}_{t=-x}$ is tabulated.

In using, for example (24), proceed thus: Multiply $8\tau^8$ by $\theta$ and add $7\tau^7$; multiply by $\theta$ and add $6\tau^6$; continue in this way until $\tau$ has been added. The required derivative is then obtained by multiplication by $1/h$ This is the process indicated when (24) is written in the form

$$hf'(x+\theta h) = \tau + \theta\{2\tau^2 + \theta\{3\tau^3 + \theta\{4\tau^4 + \ldots\}\}\}$$

**3·3. *Numerical Examples*.** To find

$$\mathrm{Ai}'(1\cdot97) + \sqrt{3}\,\mathrm{Bi}'(1\cdot97) = 2G(1\cdot97)\sin\{\psi(1\cdot97)+60°\} \quad\text{and}\quad \mathrm{Bi}'(-2\cdot57)$$

$\mathrm{Ai}'(1\cdot97) = -0\cdot05521\,805$ is taken directly from Table I. $\mathrm{Bi}'(1\cdot97)$ is obtained from Table IV, using (i) $x = 2\cdot0$, $\theta = -0\cdot3$ and (ii) $x = 1\cdot9$, $\theta = 0\cdot7$ in (24). Individual terms are set out below, and a third column (iii) shows the application of (26); the comma indicates an extra decimal retained to minimize accumulation of error.

|  | (i) $\mathrm{Bi}'(1\cdot97)$ | (ii) $\mathrm{Bi}'(1\cdot97)$ |  | (iii) $\mathrm{Bi}'(-2\cdot57)$ |
|---|---|---|---|---|
|  | $x = 2\cdot0,\ \theta = -0\cdot3$ | $x = 1\cdot9,\ \theta = +0\cdot7$ |  | $x = 2\cdot5,\ \theta = +0\cdot7$ |
| $\tau$ | $+0\cdot41006\,8205$ | $+0\cdot34951\,6586$ | $(-\tau)$ | $-0\cdot02204\,2015$ |
| $2\theta\tau^2$ | $-\quad 1978\,8570{,}0$ | $+\quad 3882\,5052{,}3$ | $-2\theta\tau^2$ | $-\quad 756\,7393{,}3$ |
| $3\theta^2\tau^3$ | $+\quad 51\,7475{,}7$ | $+\quad 234\,2198{,}0$ | $3\theta^2(-\tau^3)$ | $+\quad 2\,9063{,}8$ |
| $4\theta^3\tau^4$ | $-\quad 9627{,}2$ | $+\quad 10\,0205{,}0$ | $-4\theta^3\tau^4$ | $+\quad 1\,7970{,}2$ |
| $5\theta^4\tau^5$ | $+\quad 144{,}4$ | $+\quad 3481{,}8$ | $5\theta^4(-\tau^5)$ | $+\quad 294{,}7$ |
| $6\theta^5\tau^6$ | $-\quad 1{,}8$ | $+\quad 100{,}2$ | $-6\theta^5\tau^6$ | $-\quad 11{,}7$ |
| $7\theta^6\tau^7$ |  | $+\quad 2{,}6$ | $7\theta^6(-\tau^7)$ | $-\quad 4$ |
| $8\theta^7\tau^8$ |  | $+\quad 1$ |  |  |
| Sum | $0\cdot39078\,7626{,}1$ | $0\cdot39078\,7626{,}0$ | Sum | $-0\cdot02956\,2091{,}7$ |
| Thus | $\mathrm{Bi}'(1\cdot97) =$ | $3\cdot90787\,626$ | $\mathrm{Bi}'(-2\cdot57) =$ | $-0\cdot29562\,092$ |

It follows that

$$\mathrm{Ai}'(1\cdot97) + \sqrt{3}\,\mathrm{Bi}'(1\cdot97) = -0\cdot05521\,805 + 6\cdot76864\,023 = +\mathbf{6\cdot71342\,218}$$

the eighth decimal being unreliable to the extent of 1 or 2 units.

$G(1\cdot97)$ and $\psi(1\cdot97)$ are obtained from Table VII, using (14) and (15) respectively, with $x = 1\cdot9$ and $\theta = 0\cdot7$.

| $G(1\cdot97)$ |  | $\psi(1\cdot97)$ |  |
|---|---|---|---|
| $0\cdot3G(1\cdot9)$ | $+1\cdot04870\,64{,}9$ | $0\cdot3\psi(1\cdot9)$ | $-0°\cdot29718\,9{,}9$ |
| $0\cdot7G(2\cdot0)$ | $+2\cdot87071\,79{,}9$ | $0\cdot7\psi(2\cdot0)$ | $-0\cdot51922\,5{,}7$ |
| $E_0^2\delta_{m0}^2$ | $-\quad 433\,77{,}2$ | $E_0^2\delta_{m0}^2$ | $+\quad 339\,7{,}5$ |
| $E_1^2\delta_{m1}^2$ | $-\quad 682\,00{,}9$ | $E_1^2\delta_{m1}^2$ | $+\quad 348\,6{,}5$ |
| $M_0^4\gamma_0^4$ | $+\quad 18{,}3$ | $T^4(\gamma_0^4+\gamma_1^4)$ | $+\quad 0{,}0$ |
| $M_1^4\gamma_1^4$ | $-\quad 21{,}8$ |  |  |
| $G(1\cdot97) =$ | $3\cdot90826\,63{,}2$ | $\psi(1\cdot97) =$ | $-0°\cdot80953\,1{,}6$ |
| $2G =$ | $7\cdot81653\,26{,}4$ | $\psi+60° =$ | $59°\cdot19046\,8{,}4$ |

Thus $\quad \sin(\psi+60°) = 0\cdot85887\,470{,}7 \quad$ and $\quad 2G\sin(\psi+60°) = \mathbf{6\cdot71342\,22}$

## 4. *Definitions and Properties of the Functions*

**4·1. *Definitions of* $\mathrm{Ai}(x)$ *and* $\mathrm{Bi}(x)$.** The most convenient starting point appears to be the solution of (2) by means of a Laplace contour integral, with complex variable, using the method described by Ince in his *Ordinary Differential Equations*, p. 187, 1927. This gives

$$y = \int_C \exp\left(\tfrac{1}{3}t^3 - xt\right) dt \qquad (27)$$

where $C$ is an open contour such that the integrand vanishes at both ends. These ends must clearly be where $\tfrac{1}{3}t^3$ has infinitely negative real part, i.e. with phase between limits $(4n+1)\,\pi/6$ and $(4n+3)\,\pi/6$ for any integer $n$. Each end of $C$ must thus lie at infinity in one of the sectors of the $t$-plane that are shaded in Fig. 1. A contour $C$ beginning in the sector numbered $r$ and ending in that numbered $s$ may be denoted by $L_{rs}$. By Cauchy's theorem all contours $L_{rs}$ for given $r$ and $s$ are equivalent, since the integrand of (27) has no singularity in the finite part of the plane.

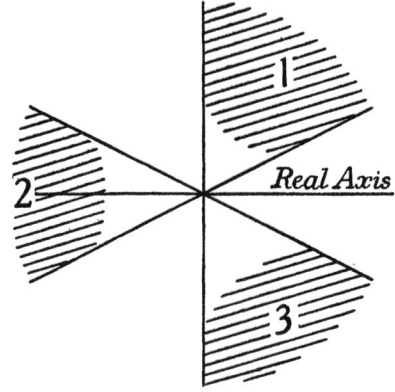

Fig. 1.

Two combinations which give real values of $y$ may be taken as independent solutions of (2); these are

$$\text{Ai}(x) = \frac{1}{2\pi i} \int_{L_{31}} \exp\left(\tfrac{1}{3}t^3 - xt\right) dt \tag{28}$$

$$\text{Bi}(x) = \frac{1}{2\pi} \left\{ \int_{L_{21}} \exp\left(\tfrac{1}{3}t^3 - xt\right) dt + \int_{L_{23}} \exp\left(\tfrac{1}{3}t^3 - xt\right) dt \right\} \tag{29}$$

To obtain from these the real integrals given on page B 17, take

$$L_{31}: \quad t = c + iu, \quad -\infty < u < \infty, \quad u \text{ real,}$$

$$L_{21}, L_{23}: \quad -\infty < t \leqslant c, \quad t \text{ real}; \quad t = c \pm iu, \quad 0 \leqslant u < \infty, \quad u \text{ real.}$$

In these $c$ is a real positive constant, which we ultimately make tend to zero.

Certain useful relations may be derived from (28) and (29) by means of the substitutions $t = \omega u$ and $t = \omega^2 u$, where $\omega = e^{2\pi i/3}$, a cube root of unity. Thus, for example,

$$\left.\begin{array}{l} \text{Ai}(x) + \omega\text{Ai}(\omega x) + \omega^2\text{Ai}(\omega^2 x) = 0 \\ \text{Bi}(x) + \omega\text{Bi}(\omega x) + \omega^2\text{Bi}(\omega^2 x) = 0 \end{array}\right\} \tag{30}$$

$$\text{Bi}(x) = i\{\omega^2\text{Ai}(\omega^2 x) - \omega\text{Ai}(\omega x)\} \tag{31}$$

It follows that

$$\left.\begin{array}{l} \text{Ai}(\omega x) = -\tfrac{1}{2}\omega^2\{\text{Ai}(x) - i\text{Bi}(x)\} \\ \text{Ai}(\omega^2 x) = -\tfrac{1}{2}\omega\{\text{Ai}(x) + i\text{Bi}(x)\} \end{array}\right\} \tag{32}$$

and

may be obtained from the tables.

4·2. The solution in series of ascending powers of $x$ by usual methods gives the general solution of (2) in the form

$$y = ay_1 + by_2 \tag{33}$$

where $y_1$ and $y_2$ are the series given on page B 17. Also, straight line contours bisecting the shaded sectors (Fig. 1) give

$$\left.\begin{array}{l} \text{Ai}(0) = 3^{-1/2}\text{Bi}(0) = 3^{-2/3}/(-\tfrac{1}{3})! = 0\cdot35502\ 80538\ 87817 \\ -\text{Ai}'(0) = 3^{-1/2}\text{Bi}'(0) = 3^{-1/3}/(-\tfrac{2}{3})! = 0\cdot25881\ 94037\ 92807 \end{array}\right\} \tag{34}$$

4·3. By comparison of expansions in ascending powers of $x$ the representations of $\text{Ai}(x)$ and $\text{Bi}(x)$ and of their derivatives in terms of Bessel functions of order $\pm 1/3$ and $\pm 2/3$ are readily derived. They are given on page B 17. Since Jeffreys (1942) remarks that "Bessel functions of order $1/3$ seem to have no application except to provide an inconvenient way of expressing this function", i.e. the function $\text{Ai}(x)$, inverted relations are given below, with $x = (\tfrac{3}{2}\xi)^{2/3}$.

$$\left.\begin{array}{ll} J_{1/3}(\xi) = \tfrac{1}{2}x^{-1/2}\{3\text{Ai}(-x) - \sqrt{3}\,\text{Bi}(-x)\} & I_{1/3}(\xi) = \tfrac{1}{2}x^{-1/2}\{\sqrt{3}\,\text{Bi}(x) - 3\text{Ai}(x)\} \\ J_{-1/3}(\xi) = \tfrac{1}{2}x^{-1/2}\{3\text{Ai}(-x) + \sqrt{3}\,\text{Bi}(-x)\} & I_{-1/3}(\xi) = \tfrac{1}{2}x^{-1/2}\{\sqrt{3}\,\text{Bi}(x) + 3\text{Ai}(x)\} \\ J_{2/3}(\xi) = \tfrac{1}{2}x^{-1}\{\sqrt{3}\,\text{Bi}'(-x) + 3\text{Ai}'(-x)\} & I_{2/3}(\xi) = \tfrac{1}{2}x^{-1}\{\sqrt{3}\,\text{Bi}'(x) + 3\text{Ai}'(x)\} \\ J_{-2/3}(\xi) = \tfrac{1}{2}x^{-1}\{\sqrt{3}\,\text{Bi}'(-x) - 3\text{Ai}'(-x)\} & I_{-2/3}(\xi) = \tfrac{1}{2}x^{-1}\{\sqrt{3}\,\text{Bi}'(x) - 3\text{Ai}'(x)\} \\ \quad\quad K_{1/3}(\xi) = \sqrt{3}\,\pi x^{-1/2}\,\text{Ai}(x) & \quad\quad K_{2/3}(\xi) = -\sqrt{3}\,\pi x^{-1}\,\text{Ai}'(x) \end{array}\right\} \tag{35}$$

4·4. The asymptotic expansions have been derived by the method of steepest descents. The contours $L_{rs}$ in (28) and (29) are chosen to pass through the saddle points of the integrand, in such directions that the modulus may fall away as rapidly as possible from its maximum values. The saddle points are given by $t = \pm \sqrt{x}$ and the directions of steepest descent to and from these points are illustrated in Figs. 2 and 3 for the two cases $x > 0$ and $x < 0$.

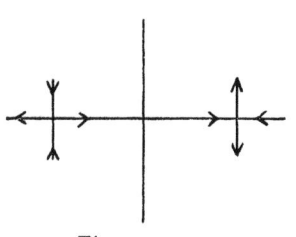

Fig. 2. $x > 0$.

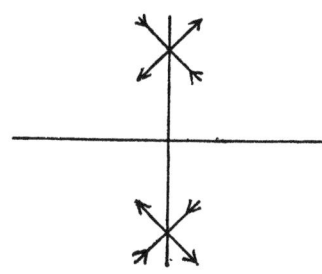

Fig. 3. $x < 0$.

Lack of space forbids the discussion of any but the simplest case, that of $\mathrm{Ai}(x)$ with $x > 0$. For this take

$$L_{31}: \qquad t = +\sqrt{x} + iu, \quad u \text{ real}, \quad -\infty < u < \infty. \tag{36}$$

Equation (28) now gives

$$\pi \exp\left(\tfrac{2}{3}x^{3/2}\right) \mathrm{Ai}(x) = \int_0^\infty e^{-u^2\sqrt{x}} \cos \tfrac{1}{3}u^3 du = \tfrac{1}{2}x^{-1/4} \int_{-\infty}^\infty e^{-v^2} \cos\left(\tfrac{1}{3}v^3 x^{-3/4}\right) dv$$

$$= \tfrac{1}{2}x^{-1/4} \int_{-\infty}^\infty e^{-v^2}\left(1 - \frac{v^6}{2!\,3^2 x^{3/2}} + \frac{v^{12}}{4!\,3^4 x^3} - \dots\right) dv$$

$$\sim \tfrac{1}{2}\pi^{1/2}x^{-1/4}\left(1 - \frac{3\cdot 5}{1!\,144 x^{3/2}} + \frac{5\cdot 7\cdot 9\cdot 11}{2!\,144^2 x^3} - \dots\right) \tag{37}$$

The other asymptotic expansions given on page B 17 may be derived similarly, although there are some troublesome variations which demand care. The asymptotic expansions for $\mathrm{Ai}(-x)$ and $\mathrm{Bi}(-x)$ were used to suggest a suitable ratio for the constants in the definitions (28) and (29).

4·5. *The Auxiliary Functions $F(x)$, $\chi(x)$, $G(x)$ and $\psi(x)$.* Combination of (2) and (4) gives

$$(F'' - F\chi'^2 - xF) \sin(\chi + \epsilon) + (2F'\chi' + F\chi'') \cos(\chi + \epsilon) = 0 \tag{38}$$

where, for brevity, the argument $(x)$ has been dropped. This must be satisfied for all $\epsilon$ so that the coefficients of $\sin(\chi + \epsilon)$ and $\cos(\chi + \epsilon)$ must vanish separately. The vanishing of the latter coefficient leads to

$$F^2\chi' = \text{const.} = -1/\pi \tag{39}$$

where the value of the constant is determined by (6), which gives

$$z \equiv \{F(x)\}^2 = \mathrm{Ai}^2(x) + \mathrm{Bi}^2(x) \qquad\qquad \tan \chi(x) = \mathrm{Ai}(x)/\mathrm{Bi}(x) \tag{40}$$

Eliminating $\chi'$, (38) and (39) now give

$$F'' - 1/\pi^2 F^3 = xF \tag{41}$$

or, in terms of $z$, after a further differentiation,

$$z''' - 4xz' - 2z = 0 \tag{42}$$

which is a linear equation. It may be verified that the complete solution of this last equation is

$$z = a\mathrm{Ai}^2(x) + b\mathrm{Ai}(x)\,\mathrm{Bi}(x) + c\mathrm{Bi}^2(x) \tag{43}$$

From (42) the asymptotic expansion for $\{F(x)\}^2$ on page B 48 may be derived. Some rather heavy algebra and use of (39) then leads to the expansion given for $\chi(x)$.

4·6. The expansions for $\{G(x)\}^2$ and $\psi(x)$ may be derived similarly from the equation

$$x\eta'' = \eta' + x^2\eta \tag{44}$$

satisfied by

$$\eta \equiv y' = A\,\mathrm{Ai}'(x) + B\,\mathrm{Bi}'(x) \tag{45}$$

or, more easily, by differentiating the relations (6). Thus

$$G^2\psi' = x/\pi \tag{46}$$

and

$$G^2 = F'^2 + F^2\chi'^2 = F'^2 + 1/\pi^2 F^2 \tag{47}$$

Also, since $\mathrm{Ai}''(x) = x\mathrm{Ai}(x)$, etc.,

$$x^2 F^2 = G'^2 + G^2\psi'^2 = G'^2 + x^2/\pi^2 G^2 \tag{48}$$

and, again,

$$GG' = x\{\mathrm{Ai}(x)\,\mathrm{Ai}'(x) + \mathrm{Bi}(x)\,\mathrm{Bi}'(x)\} = xFF' \tag{49}$$

This enables the asymptotic expansion for $\{G(x)\}^2$ to be derived readily from that for $\{F(x)\}^2$ and leads somewhat laboriously to that for $\psi(x)$.

4·7. *Formulae for Zeros and Turning-Values.* The relations (4) and (7) provide a good approach to the determination of zeros and turning-values of the general solution of (2) and of its derivative. Zeros $c$, $c'$, of $y$ and $y'$ respectively, satisfy equations

$$\chi(c) = s\pi - \epsilon \qquad\qquad \psi(c') = s\pi - \epsilon \tag{50}$$

Reversion of the series obtained by substituting for $\chi(c)$ and $\psi(c')$ their asymptotic expansions leads to the expressions (see also page B 48)

$$c \sim -\lambda^{2/3}\left(1 + \frac{5}{48}\frac{1}{\lambda^2} - \frac{5}{36}\frac{1}{\lambda^4} + \ldots\right) \qquad \lambda = \tfrac{3}{8}\pi(4s - 1) - \tfrac{3}{2}\epsilon$$
$$c' \sim -\mu^{2/3}\left(1 - \frac{7}{48}\frac{1}{\mu^2} + \frac{35}{288}\frac{1}{\mu^4} - \ldots\right) \qquad \mu = \tfrac{3}{8}\pi(4s + 1) - \tfrac{3}{2}\epsilon \tag{51}$$

where $s$ is an integer, usually positive, but which may be zero for $c'$ if $\epsilon$ (assumed positive) is between o and $\pi/6$.

Again, since by (4) and (7)

$$y'/C = G\sin(\psi + \epsilon) = F'\sin(\chi + \epsilon) + F\chi'\cos(\chi + \epsilon)$$
$$xy/C = xF\sin(\chi + \epsilon) = G'\sin(\psi + \epsilon) + G\psi'\cos(\psi + \epsilon) \tag{52}$$

it follows that

$$y'(c) = \pm CF\chi' = \mp C/\pi F(c)$$
$$y(c) = \pm CG\psi'/x = \pm C/\pi G(c') \tag{53}$$

giving the turning-values of $y'$ and $y$.

4·8. Asymptotic expansions, such as (51), are satisfactory only when the zero is large, but an alternative possibility, that of inverse interpolation into Table VII, is available in many cases.

Values of $y$ and $y'$ (e.g., obtained from (3) or (45)) may also be used as a basis for interpolation, inverse or direct, to give zeros and turning-values. The inverse interpolation may be troublesome, but can be avoided as follows:

Suppose that $x = k$ is an approximation to a zero $c$ of a solution $y$ of (2). Write

$$c = k + h \tag{54}$$

Now it is known that an approximation to the value of $h$ is $-y(k)/y'(k)$; write, therefore,

$$u(k) = y(k)/y'(k) \tag{55}$$

and develop $h$ as a power series in $u$ and $k$. Then starting with any suitable value of $k$ the corresponding value of $u$ is readily obtained and the value of $c$ may be calculated; it is important that the unknown constant $c$ should not occur explicitly in the expansion for $h$.

Total differentiation of (54) with respect to $k$, remembering that $h$ is an explicit function of $k$ and $u$, while $u$ is itself a function of $k$, gives

$$o = 1 + \frac{\partial h}{\partial k} + \frac{\partial h}{\partial u}\cdot\frac{du}{dk} \tag{56}$$

But from (55), using (2),

$$du/dk = 1 - ku^2 \tag{57}$$

whence

$$1 + \frac{\partial h}{\partial k} + (1 - ku^2)\frac{\partial h}{\partial u} = o \tag{58}$$

Substituting

$$h = -u + a_2u^2 + a_3u^3 + \ldots + a_nu^n + \ldots$$

where $a_n$ may be a function of $k$, in (58) then leads (see page B 48) to

$$c = k - u - 2ku^3/3! + 2u^4/4! - 24k^2u^5/5! + \ldots \tag{59}$$

In a similar fashion, writing

$$y'(c)/y'(k) = 1 + \lambda \tag{60}$$

leads to the equation

$$(1 + \lambda)ku + \frac{\partial\lambda}{\partial k} + (1 - ku^2)\frac{\partial\lambda}{\partial u} = o \tag{61}$$

whence

$$y'(c) = y'(k)(1 - ku^2/2! + u^3/3! - 3k^2u^4/4! + 14ku^5/5! - \ldots) \tag{62}$$

Series for $c'$ and $y(c')/y(k')$ have been derived similarly, and are given on page B 48.

4·9. *Numerical Application of* (59) *and* (62). To find the zero of $y = \mathrm{Ai}(x) - \mathrm{Bi}(x)$ near $x = k = -0.4$.

|  | $c$ | $y'(c)/y'(k)$ |
|---|---|---|
| $y(-0.4) = +0.02420\ 467$ | $-0.40000\ 000$ | $+1.00000\ 000$ |
| $y'(-0.4) = -0.71276\ 627$ | $+\ \ \cdot 03395\ 877,6$ | $+0.00023\ 064,0$ |
| $u(-0.4) = -0.03395\ 877,6$ | $-\ \ \ \ \ \ \ \ \ 522,1$ | $-\ \ \ \ \ \ \ \ \ 652,7$ |
|  | $+\ \ \ \ \ \ \ \ \ \ \ 11,1$ | $-\ \ \ \ \ \ \ \ \ \ \ \ 2,7$ |
|  | $+\ \ \ \ \ \ \ \ \ \ \ \ \ 1$ | $+\ \ \ \ \ \ \ \ \ \ \ \ \ 2$ |
|  | Sum $-0.36604\ 633,3$ | $+1.00022\ 408,8$ |

Thus $\qquad c = -0.36604\ 633(3)$ and $y'(c) = -0.71292\ 599(2)$

From $k = -0.3$ $\qquad c = -0.36604\ 632(1)$ and $y'(c) = -0.71292\ 600(2)$

## 5. *Preparation of the Tables*

5·1. *Computation of Pivotal Values.* Basic or pivotal values of all functions were first calculated to at least three more decimals than are given in the final tables. The interval between successive arguments was chosen so that intermediate values could be found to the same accuracy, using not more than 8 or 10 terms of an appropriate interpolation formula. The required intermediate values for the functions $\mathrm{Ai}(x)$ and $\mathrm{Ai}'(x)$ were then derived to 10 decimals by subtabulation, using the Association's National machine, and the values in Table I were obtained. Apart from $\mathrm{Ai}'(x)/\mathrm{Ai}(x)$ with $20 \leqslant x \leqslant 25$ and the auxiliary functions $F$, $\chi$, $G$ and $\psi$ with $-30 \leqslant x \leqslant -10$, for which subtabulation from unit interval was found to be practicable, other functions were obtained directly without subtabulation.

The calculation of the pivotal values of the various functions led the writer to make various investigations of method; it seems desirable to indicate here only those which have been found most effective. The main method used for obtaining these pivotal values was step-by-step application of the Taylor expansions (23) and (24) with $\theta = \pm 1$, the appropriate differential equation being used to give $\tau^n$ from $f(x)$ and $f'(x)$. As a rule $h$ was taken as $0.1$, although $h = 0.05$ was found to be desirable for large values of $-x$.

For $\mathrm{Ai}(x)$ and $\mathrm{Bi}(x)$ repeated differentiation of (2) gives

$$(n + 1)(n + 2)\tau^{n+2} = h^2(x\tau^n + h\tau^{n-1}) \qquad n > 0$$

$$2\tau^2 = h^2 xy$$

These were found to give a most effective and rapid method of calculation. Further, the method is self-checking. Each step gives $f(x \pm h)$ and $f'(x \pm h)$, one of each pair of values is new, the other a check reproduction of a previous value. The effect of accumulation of error is discussed in § 5·2; it was found to be negligible.

The series for $\mathrm{Ai}(x)$ and $\mathrm{Bi}(x)$ in ascending powers of $x$ and the asymptotic expansions (see page B 17) were used to give additional check values as follows:

$$\mathrm{Ai}(x), \quad \text{for} \quad \pm x = 0.1(0.1)\ 1, 2, 5, 10, 20; \qquad \mathrm{Bi}(x), \quad \text{for} \quad x = 1, 2, 5, 10. \qquad (63)$$

For $\mathrm{Ai}'(x)/\mathrm{Ai}(x)$ in Table II and for $\mathrm{Bi}'(x)/\mathrm{Bi}(x)$ in Table VI the appropriate differential equation is

$$z' = x - z^2 \qquad (64)$$

where $z = y'/y$, $y$ being a solution of (2). This equation was used for $x \leqslant 20$. The values of $\log_{10}\mathrm{Ai}(x)$ and of $\log_{10}\mathrm{Bi}(x)$ were then obtained by numerical integration of $z$, partly by use of the Taylor expansion (22) with $\theta = 1$, and partly by use of the formula

$$h\int_0^1 f(x + \theta h)\,d\theta = \tfrac{1}{2}h\{(f_0 + f_1) - \tfrac{1}{12}(\delta_0^2 + \delta_1^2) + \tfrac{11}{720}(\delta_0^4 + \delta_1^4) - \tfrac{191}{60480}(\delta_0^6 + \delta_1^6) + \ldots\} \qquad (65)$$

which was applied, with $h = 0.1$, in the form

$$\Delta \log_{10}\mathrm{Ai}(x) = Z_0 + Z_1 \qquad Z_r = \frac{M}{14400}(720 - 60\delta^2 + 11\delta^4 - 2.27\delta^6 + \ldots)z_r \qquad (66)$$

where $M = \log_{10} e$. For $x \geqslant 20$, asymptotic expansions for $\mathrm{Ai}'(x)/\mathrm{Ai}(x)$ and $\log_{10}\mathrm{Ai}(x)$ were used at unit interval in $x$. These expansions were derived from that for $\mathrm{Ai}(x)$, (page B 17).

For the auxiliary function $F$ equation (41) was used. As this is not convenient for repeated differentiation, the method described in *Interpolation and Allied Tables* (page 942, 1942 reprint, not in the 1936 edition; London, H.M. Stationery Office) was employed; this method, which involves (central) sums and differences instead of derivatives, is very convenient when the first derivative is absent from the differential equation. The numerical integration of the equation for $F$ was carried out for $x = -11 \cdot 0 (0 \cdot 1) + 2 \cdot 8$. Numerical quadrature, using (39) and (65) then gave $\chi$ in radians, and the values were converted to degrees.

$G$ was calculated from (47) using values of $F'$ derived from the tables of sums and differences obtained when computing $F$. Finally $\psi$ was obtained from (46) in the same way that $\chi$ was obtained from (39). The relations (6) and (8) provided useful checks at various points.

For each of the auxiliary functions when $x \leqslant -9$, the appropriate asymptotic expansion (page B48) was used, at unit interval in $x$.

Zeros were calculated from the asymptotic expansions wherever they gave sufficient accuracy. Within the tabular range of Ai$(x)$ and Bi$(x)$ zeros were calculated by successive approximations, using reduced derivatives in Taylor expansions (this work was completed before the more efficient expansions such as (59) were discovered). The overlap of the two sets of results provided valuable checks.

5·2. *Checking of Pivotal Values.* All basic or pivotal values were checked by differencing on the Association's National machine. Wherever this check was inadequate, as, for example, with the earlier zeros, values were recalculated by an alternative route; often three distinct calculations were made in these regions.

When a function is calculated by the self-checking Taylor series method, or by the method used for $F$, the main possibility of error arises from accumulation; such errors cannot be detected by differencing. The effect of these errors was minimized by performing the step-by-step integration in the appropriate direction; for Bi$(x)$ this is always away from the origin, for Ai$(x)$ always towards $x = -\infty$.

To illustrate this point, consider, for example, Ai$(x)$; the solution (3) to the equation (2) should then have $A = 1$, $B = 0$. At each step, however, a slight error must be made (numerical calculation being necessarily limited in accuracy) and we shall have $A = 1 + \epsilon_1$, $B = \epsilon_2$, where $\epsilon_1$ and $\epsilon_2$ are two "accidental" functions of $x$ which vary irregularly but whose amount is required to remain small. If we follow up a single pair of errors, however, we shall have $\epsilon_1$ and $\epsilon_2$ constant. Reference to the graph of Ai$(x)$ and Bi$(x)$ on page B16 shows that as $x$ ($> 0$) increases towards large positive $x$, the error term involving $\epsilon_2$ eventually predominates, even over Ai$(x)$, however small $\epsilon_2$ ($\neq 0$) may be. On the other hand, working towards the origin from the positive side, the $\epsilon_2$ term dies away, while the $\epsilon_1$ term bears a constant ratio to Ai$(x)$.

Consider next $x < 0$. The errors in Ai$(x)$ and Ai$'(x)$ may be written

$$F \sqrt{\epsilon_1^2 + \epsilon_2^2} \sin (\chi - \alpha) \qquad\qquad G \sqrt{\epsilon_1^2 + \epsilon_2^2} \sin (\psi - \alpha) \qquad\qquad (67)$$

with $\tan \alpha = \epsilon_2/\epsilon_1$, and if suffix zero be used to indicate the point of origin of the particular error considered, then

$$F_0 \sqrt{\epsilon_1^2 + \epsilon_2^2} \sin (\chi_0 - \alpha) < K \qquad\qquad G_0 \sqrt{\epsilon_1^2 + \epsilon_2^2} \sin (\psi_0 - \alpha) < K' \qquad\qquad (68)$$

where $K$ and $K'$ depend only on the precision of the calculation at $x = x_0$ (for Ai$(x)$ and Bi$(x)$, $K$ and $K'$ were both of order $10^{-11}$ or $10^{-12}$). Now $|\chi - \psi| \geqslant 60°$ for $x \leqslant 0$, so that one at least of $\chi_0 - \alpha$, $\psi_0 - \alpha$ differs by more than $30°$ from the nearest multiple of $180°$. Hence $(\epsilon_1^2 + \epsilon_2^2)^{1/2}$ cannot exceed the larger of $2K/F_0$ and $2K'/G_0$; the errors in Ai$(x)$ and Ai$'(x)$ are thus bounded by moderate multiples of $K$ or $K'$ throughout the range of the integration, since for $-20 \leqslant x \leqslant 0$ no multiple of $K$ or $K'$ can exceed $2G(-20)/F(-20) < 10$. Since $F$ decreases as $-x \to +\infty$, the integration was performed outward from $x = 0$; the reverse sense, towards $x = 0$, would be almost equally good when $x < 0$ —perhaps better for the derivative.

Thus the cumulative error is composed of a number of limited errors, with limits of comparable size, and these may be expected to accumulate very nearly according to the normal or Gaussian law. It is consequently sufficient to examine the error at fairly widely separated intervals in order to study the accumulation. The isolated calculations at the values of $x$ given in (63) were found to be ample for this purpose. The final conclusion is that the pivotal values are all accurate to well within a unit of the second decimal beyond the last given in the tables.

5·3. *Subtabulation.* The methods of subtabulation used were very similar to those described by Comrie ("Inverse Interpolation" and "Scientific Applications of the National Accounting Machine", *Journal of the Royal Statistical Society* (Supplement) **3**, 87–114, 1936). Owing to the use of rather wide intervals, certain extensions were necessary, but the main features remain; the subtabulation was done on the National machine in such a way that an error automatically resulted in subsequent failure to reproduce pivotal values. The subtabulation carried two, and sometimes three, decimals more than are printed; where only two extra decimals were kept (in the preparation of Table I) all borderline cases, where correct rounding off to 8 decimals was in doubt, were separately examined, partly because, at that stage, the effect of cumulative error had not been investigated (see § 5·2).

5·4. *Preparation of Printer's Copy.* Wherever a table was produced by subtabulation (i.e. Table I and parts of Tables II and VII), the printing of the sheets during the final stage on the National machine was arranged in such a way that three columns containing argument, function and second difference could be detached and used for printer's copy.

Table VI and the remainder of Tables II and VII consist of pivotal values, together with appropriate differences. These have all been checked by differencing on the National machine, this process either (*a*) supplied detachable printer's copy or (*b*) was applied to the proof; in this way copying errors were eliminated.

Where $\delta_m^2$ is needed, the values of $\delta^2$ were adjusted by hand. Two separate calculations of the modified differences were made, using (16), usually by different computers; wherever this was not done, the modified differences were checked in proof by forming $\delta^2 - \delta_m^2$ on the National machine and comparing with the product $0\cdot184\delta^4$. When forming modified differences the order of (i) the application of (16) and (ii) the rounding-off of the function and of the differences to tabular accuracy was not always the same; the more convenient order was used in each case. Alteration in the order of application of these two processes may cause a maximum change of 3 units, very rarely attained, in $\delta_m^2$, but the effect on an interpolated value is negligible, since a large change in one direction is always associated with substantial changes, in the opposite direction, of the adjacent differences.

Values of $\gamma^4$ were entered on the copy by hand, after determination by (16), use of which always preceded rounding off in these cases; duplicate calculations were used in checking.

The printer's copy for the remaining Tables III, IV and V was made in typescript.

5·5. *Checking of Proofs.* The printer's copy for Tables I, II, VI and VII, having been prepared on the National machine, could be considered free from copying errors (except possibly in the argument columns, and in signs, headings, etc.); the typed Tables III, IV and V on the other hand, were not assumed to be entirely free from typing errors. This variation was allowed for when reading proofs. An outline of the minimum procedure accepted as satisfactory is given below, in places even more readings were made.

First proofs were compared twice with printer's copy in the case of the former set of tables (for which this copy forms the *only* complete original), and once in the case of the latter set of tables, which was also compared with the original calculations. All printed pivotal values were also carefully compared with the original calculations, and all end-figures, leading figures, signs, arguments, headings, etc., were independently checked. Revised proofs were read completely against original calculations. Since the printing was done from type—and not from stereo plates—the final copy from the press after the main printing of the tables was again read against the printer's copy.

It is hoped and expected, therefore, that the tabular values given are all correct within half a unit of the final figure printed; also that the differences are correct, subject to their derivation from the tabular values, and to the possible variation in $\delta_m^2$ indicated in §5·4.

Special methods were needed for the pages of formulae. These have all been read twice against carefully prepared copy, and against the original work. Where possible the formulae were also used, in proof, in typical check calculations.

## 6. *Acknowledgements*

The writer's thanks are due to Dr Harold Jeffreys, who originally suggested to the Committee the tabulation of the Airy Integral, for his continued encouragement and helpful suggestions, and for the background of theory which he provided; to Dr L. J. Comrie who, in 1933, indicated the first stretches of the path which led to the present tables, and who has continued to take great interest in the work and to give much valuable advice, and to Dr F. W. Bradley, who provided valuable help with theoretical investigations. Much help was given by the Committee and in particular by Dr W. G. Bickley and Dr A. J. Thompson who, with the writer, constituted a sub-committee on these tables, and by Professor E. H. Neville; this help is gratefully acknowledged.

Very valuable help with the computations connected with the second solution Bi$(x)$ and with the zeros of both functions and of their derivatives was given by Mr C. E. Gwyther, an indefatigable and extraordinarily accurate worker on the Committee's behalf, and by Mr T. J. Lunt. The difficult sub-tabulations were ably performed by Dr H. O. Hartley and Mrs R. O. Cashen. The block for the graph on page B16 was kindly lent by Dr L. J. Comrie. To all these and to other Committee computers the writer is greatly indebted. Assistance with the proof reading was given by Dr A. Fletcher, by Mr C. E. Gwyther and by Mr C. W. Jones; this is also gratefully acknowledged.

Finally, thanks are due to Professor L. Rosenhead and to the Liverpool University Authorities for placing at the writer's disposal the equipment of the Mathematical Laboratory during various stages of the computations, and to the Staff of the Cambridge University Press for their co-operation during the printing of the tables.

J. C. P. MILLER

# BIBLIOGRAPHY

AIRY, SIR G. B. 1838. On the Intensity of Light in the neighbourhood of a Caustic. *Trans. Camb. Phil. Soc.* **6**, 379–402.

—— 1849. Supplement to a Paper "On the Intensity of Light in the neighbourhood of a Caustic". *Trans. Camb. Phil. Soc.* **8**, 595–599.

BRILLOUIN, L. 1916. Sur une Méthode de calcul approchée de certaines Intégrales, dite Méthode de Col. *Ann. Sci. de l'École Normale Supérieure* (3) **33**, 17–69.

JEFFREYS, H. 1924. (i) On certain approximate solutions of linear differential equations of the second order. (ii) On certain solutions of Mathieu's equation. (iii) On the modified Mathieu's equation. (iv) The free oscillations of water in an elliptical lake. *Proc. Lond. Math. Soc.* (2) **23**, 428–476.

—— 1928. The Effect on Love Waves of Heterogeneity in the Lower Layer. *Month. Not. Roy. Astr. Soc., Geoph. Supp.* **2**, 101–111.

—— 1942. Asymptotic Solutions of Linear Differential Equations. *Phil. Mag.* (7) **33**, 451–456.

KRAMERS, H. A. 1926. Wellenmechanik und halbzahlige Quantisierung. *Zs. f. Physik* **39**, 828–840.

NICHOLSON, J. W. 1909. On the Relation of Airy's Integral to the Bessel Functions. *Phil. Mag.* (6) **18**, 6–17.

RAYLEIGH, LORD. 1915. On the Stability of the Simple Shearing Motion of a Viscous Incompressible Fluid. *Phil. Mag.* (6) **30**, 329–338. Also in *Sci. Papers* **6**, 341–349, 1920.

STOKES, SIR G. G. 1851. On the numerical Calculation of a Class of Definite Integrals and Infinite Series. *Trans. Camb. Phil. Soc.* **9** (1), 166–187. Also in *Math. and Phys. Papers* **2**, 329–357, 1883.

—— 1858. On the Discontinuity of Arbitrary Constants which appear in Divergent Developments. *Trans. Camb. Phil. Soc.* **10** (1), 106–128. Also in *Math. and Phys. Papers* **4**, 77–109, 1904.

—— 1907. *Mem. and Sci. Corr.* **2**, 159–160.

SZEGÖ, G. 1939. *Orthogonal Polynomials.* New York City, American Math. Soc.

WATSON, G. N. 1922. *Treatise on the Theory of Bessel Functions.* Cambridge, University Press.

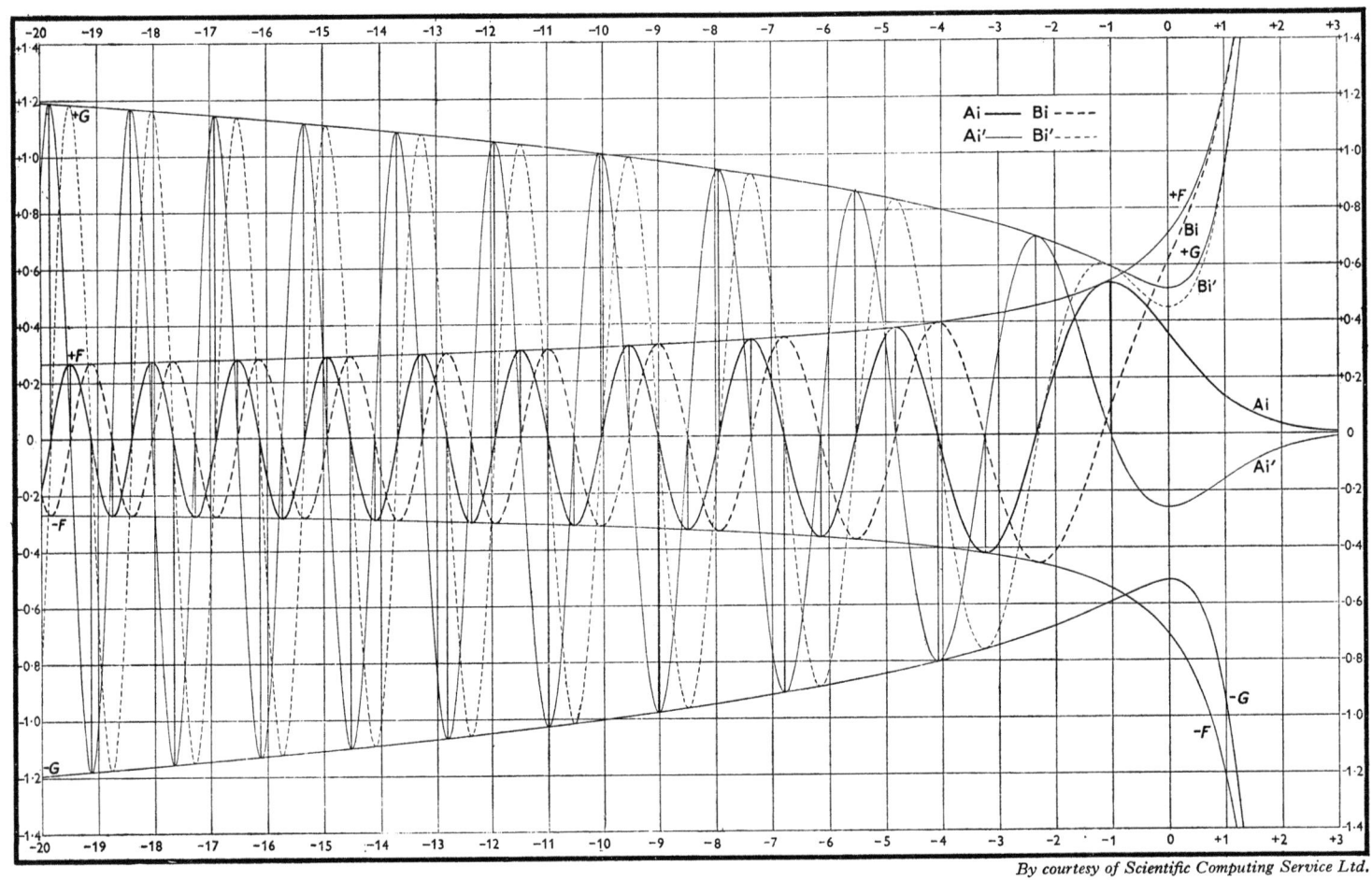

By courtesy of Scientific Computing Service Ltd.

The diagram shows $\mathrm{Ai}(x)$ and $\mathrm{Bi}(x)$, their first derivatives $\mathrm{Ai}'(x)$ and $\mathrm{Bi}'(x)$, and also $\pm F(x)$ and $\pm G(x)$, where

$$\{F(x)\}^2 = \mathrm{Ai}^2(x) + \mathrm{Bi}^2(x) \qquad \text{and} \qquad \{G(x)\}^2 = \mathrm{Ai}'^2(x) + \mathrm{Bi}'^2(x)$$

Since $\mathrm{Ai}''(x) = x\,\mathrm{Ai}(x)$, it follows that a stationary value of *either* of the functions $\mathrm{Ai}(x)$ or $\mathrm{Ai}'(x)$ corresponds to a zero of the other: this is indicated by insertion of the appropriate ordinates. A similar relation holds between $\mathrm{Bi}(x)$ and $\mathrm{Bi}'(x)$, but in this case the ordinates have not been included in the diagram.

# THE FUNCTIONS Ai(x) AND Bi(x)

*Differential equation:*

Ai(x) and Bi(x) are independent solutions of the differential equation

$$\frac{d^2y}{dx^2} = xy$$

They satisfy the identity $\quad$ Ai$(x)$ Bi$'(x) -$ Ai$'(x)$ Bi$(x) = 1/\pi$

*Integral representations:*

$$\text{Ai}(x) = \frac{1}{\pi} \int_0^\infty \cos\left(\tfrac{1}{3}t^3 + xt\right) dt$$

$$\text{Bi}(x) = \frac{1}{\pi} \int_0^\infty \left\{\exp\left(-\tfrac{1}{3}t^3 + xt\right) + \sin\left(\tfrac{1}{3}t^3 + xt\right)\right\} dt$$

*Power series:*

$$\text{Ai}(x) = \alpha y_1 - \beta y_2 \qquad\qquad \text{Bi}(x) = 3^{1/2}\left(\alpha y_1 + \beta y_2\right)$$

where $\quad y_1 = 1 + \dfrac{1}{3!}x^3 + \dfrac{1\cdot4}{6!}x^6 + \dfrac{1\cdot4\cdot7}{9!}x^9 + \ldots \qquad y_2 = x + \dfrac{2}{4!}x^4 + \dfrac{2\cdot5}{7!}x^7 + \dfrac{2\cdot5\cdot8}{10!}x^{10} + \ldots$

and $\quad \alpha = 3^{-2/3}/(-\tfrac{1}{3})! = 0.35502\ 80538\ 87817 \qquad \beta = 3^{-1/3}/(-\tfrac{2}{3})! = 0.25881\ 94037\ 92807$

*Representations in terms of Bessel functions:*

If $x$ is real and positive, and if $\xi = \tfrac{2}{3}x^{3/2}$,

$$\text{Ai}(x) = \tfrac{1}{3}x^{1/2}\left\{I_{-1/3}(\xi) - I_{1/3}(\xi)\right\} \qquad \text{Ai}(-x) = \tfrac{1}{3}x^{1/2}\left\{J_{-1/3}(\xi) + J_{1/3}(\xi)\right\}$$

$$\text{Bi}(x) = (\tfrac{1}{3}x)^{1/2}\left\{I_{-1/3}(\xi) + I_{1/3}(\xi)\right\} \qquad \text{Bi}(-x) = (\tfrac{1}{3}x)^{1/2}\left\{J_{-1/3}(\xi) - J_{1/3}(\xi)\right\}$$

$$\pi\text{Ai}(x) = (\tfrac{1}{3}x)^{1/2} K_{1/3}(\xi)$$

*Asymptotic expansions:*

If $\xi$ has the same meaning as before, then

$$\text{Ai}(x) \sim \tfrac{1}{2}\pi^{-1/2}x^{-1/4}e^{-\xi}\left(1 - \frac{3\cdot5}{1!\,216}\frac{1}{\xi} + \frac{5\cdot7\cdot9\cdot11}{2!\,216^2}\frac{1}{\xi^2} - \frac{7\cdot9\cdot11\cdot13\cdot15\cdot17}{3!\,216^3}\frac{1}{\xi^3} + \ldots\right)$$

$$\text{Bi}(x) \sim \pi^{-1/2}x^{-1/4}e^{\xi}\left(1 + \frac{3\cdot5}{1!\,216}\frac{1}{\xi} + \frac{5\cdot7\cdot9\cdot11}{2!\,216^2}\frac{1}{\xi^2} + \frac{7\cdot9\cdot11\cdot13\cdot15\cdot17}{3!\,216^3}\frac{1}{\xi^3} + \ldots\right)$$

and $\qquad \text{Ai}(-x) = \pi^{-1/2}x^{-1/4}\left\{P(x)\sin(\xi + \tfrac{1}{4}\pi) - Q(x)\cos(\xi + \tfrac{1}{4}\pi)\right\}$

$\qquad\qquad \text{Bi}(-x) = \pi^{-1/2}x^{-1/4}\left\{P(x)\cos(\xi + \tfrac{1}{4}\pi) + Q(x)\sin(\xi + \tfrac{1}{4}\pi)\right\}$

where $\qquad P(x) \sim 1 - \dfrac{5\cdot7\cdot9\cdot11}{2!\,216^2}\dfrac{1}{\xi^2} + \dfrac{9\cdot11\cdot13\cdot15\cdot17\cdot19\cdot21\cdot23}{4!\,216^4}\dfrac{1}{\xi^4} - \ldots$

$$Q(x) \sim \dfrac{3\cdot5}{1!\,216}\dfrac{1}{\xi} - \dfrac{7\cdot9\cdot11\cdot13\cdot15\cdot17}{3!\,216^3}\dfrac{1}{\xi^3} + \ldots$$

Formulae for auxiliary functions and zeros are given on page B 48. The introduction should also be consulted for further formulae.

TABLE I—Ai($x$) AND Ai$'$($x$)

| $x$ | Ai($x$) | $\delta^2$ | Ai$'$($x$) | $\delta^2$ | Ai($-x$) | $\delta^2$ | Ai$'$($-x$) | $\delta^2$ |
|---|---|---|---|---|---|---|---|---|
| 0·00 | +0·35502 805 | + 1 | −0·25881 940 | +3 549 | +0·35502 805 | + 1 | −0·25881 940 | +3 549 |
| ·01 | ·35243 992 | 35 | ·25880 174 | 3 499 | ·35761 619 | − 36 | ·25880 157 | 3 603 |
| ·02 | ·34985 214 | 69 | ·25874 909 | 3 447 | ·36020 397 | 73 | ·25874 771 | 3 654 |
| ·03 | ·34726 505 | 105 | ·25866 197 | 3 395 | ·36279 102 | 108 | ·25865 731 | 3 705 |
| ·04 | ·34467 901 | 138 | ·25854 090 | 3 343 | ·36537 699 | 147 | ·25852 986 | 3 757 |
| 0·05 | +0·34209 435 | + 170 | −0·25838 640 | +3 292 | +0·36796 149 | − 183 | −0·25836 484 | +3 809 |
| ·06 | ·33951 139 | 204 | ·25819 898 | 3 240 | ·37054 416 | 223 | ·25816 173 | 3 861 |
| ·07 | ·33693 047 | 236 | ·25797 916 | 3 189 | ·37312 460 | 261 | ·25792 001 | 3 911 |
| ·08 | ·33435 191 | 268 | ·25772 745 | 3 137 | ·37570 243 | 301 | ·25763 918 | 3 963 |
| ·09 | ·33177 603 | 298 | ·25744 437 | 3 087 | ·37827 725 | 340 | ·25731 872 | 4 015 |
| 0·10 | +0·32920 313 | + 329 | −0·25713 042 | +3 034 | +0·38084 867 | − 381 | −0·25695 811 | +4 065 |
| ·11 | ·32663 352 | 360 | ·25678 613 | 2 984 | ·38341 628 | 422 | ·25655 685 | 4 116 |
| ·12 | ·32406 751 | 388 | ·25641 200 | 2 933 | ·38597 967 | 463 | ·25611 443 | 4 168 |
| ·13 | ·32150 538 | 418 | ·25600 854 | 2 883 | ·38853 843 | 506 | ·25563 033 | 4 217 |
| ·14 | ·31894 743 | 447 | ·25557 625 | 2 831 | ·39109 213 | 546 | ·25510 406 | 4 268 |
| 0·15 | +0·31639 395 | + 474 | −0·25511 565 | +2 781 | +0·39364 037 | − 592 | −0·25453 511 | +4 319 |
| ·16 | ·31384 521 | 503 | ·25462 724 | 2 732 | ·39618 269 | 633 | ·25392 297 | 4 367 |
| ·17 | ·31130 150 | 528 | ·25411 151 | 2 680 | ·39871 868 | 678 | ·25326 716 | 4 419 |
| ·18 | ·30876 307 | 556 | ·25356 898 | 2 632 | ·40124 789 | 723 | ·25256 716 | 4 466 |
| ·19 | ·30623 020 | 582 | ·25300 013 | 2 581 | ·40376 987 | 766 | ·25182 250 | 4 517 |
| 0·20 | +0·30370 315 | + 608 | −0·25240 547 | +2 533 | +0·40628 419 | − 813 | −0·25103 267 | +4 564 |
| ·21 | ·30118 218 | 632 | ·25178 548 | 2 482 | ·40879 038 | 859 | ·25019 720 | 4 614 |
| ·22 | ·29866 753 | 657 | ·25114 067 | 2 435 | ·41128 798 | 905 | ·24931 559 | 4 661 |
| ·23 | ·29615 945 | 681 | ·25047 151 | 2 385 | ·41377 653 | 951 | ·24838 737 | 4 709 |
| ·24 | ·29365 818 | 704 | ·24977 850 | 2 338 | ·41625 557 | 1 000 | ·24741 206 | 4 756 |
| 0·25 | +0·29116 395 | + 729 | −0·24906 211 | +2 288 | +0·41872 461 | − 1 046 | −0·24638 919 | +4 804 |
| ·26 | ·28867 701 | 750 | ·24832 284 | 2 242 | ·42118 319 | 1 095 | ·24531 828 | 4 849 |
| ·27 | ·28619 757 | 773 | ·24756 115 | 2 193 | ·42363 082 | 1 144 | ·24419 888 | 4 895 |
| ·28 | ·28372 586 | 794 | ·24677 753 | 2 147 | ·42606 701 | 1 194 | ·24303 053 | 4 942 |
| ·29 | ·28126 209 | 816 | ·24597 244 | 2 099 | ·42849 126 | 1 241 | ·24181 276 | 4 986 |
| 0·30 | +0·27880 648 | + 836 | −0·24514 636 | +2 052 | +0·43090 310 | − 1 294 | −0·24054 513 | +5 031 |
| ·31 | ·27635 923 | 857 | ·24429 976 | 2 007 | ·43330 200 | 1 343 | ·23922 719 | 5 074 |
| ·32 | ·27392 055 | 877 | ·24343 309 | 1 960 | ·43568 747 | 1 394 | ·23785 851 | 5 118 |
| ·33 | ·27149 064 | 895 | ·24254 682 | 1 915 | ·43805 900 | 1 446 | ·23643 865 | 5 161 |
| ·34 | ·26906 968 | 915 | ·24164 140 | 1 868 | ·44041 607 | 1 497 | ·23496 718 | 5 203 |
| 0·35 | +0·26665 787 | + 934 | −0·24071 730 | +1 825 | +0·44275 817 | − 1 550 | −0·23344 368 | +5 245 |
| ·36 | ·26425 540 | 950 | ·23977 495 | 1 779 | ·44508 477 | 1 602 | ·23186 773 | 5 285 |
| ·37 | ·26186 243 | 970 | ·23881 481 | 1 736 | ·44739 535 | 1 656 | ·23023 893 | 5 326 |
| ·38 | ·25947 916 | 985 | ·23783 731 | 1 690 | ·44968 937 | 1 708 | ·22855 687 | 5 365 |
| ·39 | ·25710 574 | 1 003 | ·23684 291 | 1 648 | ·45196 631 | 1 764 | ·22682 116 | 5 404 |
| 0·40 | +0·25474 235 | +1 020 | −0·23583 203 | +1 603 | +0·45422 561 | − 1 816 | −0·22503 141 | +5 443 |
| ·41 | ·25238 916 | 1 033 | ·23480 512 | 1 562 | ·45646 675 | 1 871 | ·22318 723 | 5 479 |
| ·42 | ·25004 630 | 1 051 | ·23376 259 | 1 519 | ·45868 918 | 1 928 | ·22128 826 | 5 517 |
| ·43 | ·24771 395 | 1 066 | ·23270 487 | 1 476 | ·46089 233 | 1 981 | ·21933 412 | 5 551 |
| ·44 | ·24539 226 | 1 078 | ·23163 239 | 1 435 | ·46307 567 | 2 037 | ·21732 447 | 5 588 |
| 0·45 | +0·24308 135 | +1 095 | −0·23054 556 | +1 394 | +0·46523 864 | − 2 095 | −0·21525 894 | +5 620 |
| ·46 | ·24078 139 | 1 107 | ·22944 479 | 1 352 | ·46738 066 | 2 149 | ·21313 721 | 5 655 |
| ·47 | ·23849 250 | 1 121 | ·22833 050 | 1 311 | ·46950 119 | 2 207 | ·21095 893 | 5 686 |
| ·48 | ·23621 482 | 1 134 | ·22720 310 | 1 273 | ·47159 965 | 2 263 | ·20872 379 | 5 718 |
| ·49 | ·23394 848 | 1 147 | ·22606 297 | 1 231 | ·47367 548 | 2 322 | ·20643 147 | 5 748 |
| 0·50 | +0·23169 361 | +1 157 | −0·22491 053 | +1 192 | +0·47572 809 | − 2 378 | −0·20408 167 | +5 778 |

# TABLE I—Ai($x$) AND Ai'($x$)

| $x$ | Ai($x$) | $\delta^2$ | Ai'($x$) | $\delta^2$ | Ai($-x$) | $\delta^2$ | Ai'($-x$) | $\delta^2$ |
|---|---|---|---|---|---|---|---|---|
| 0·50 | +0·23169 361 | +1 157 | −0·22491 053 | +1 192 | +0·47572 809 | −2 378 | −0·20408 167 | +5 778 |
| ·51 | ·22945 031 | 1 171 | ·22374 617 | 1 154 | ·47775 692 | 2 437 | ·20167 409 | 5 805 |
| ·52 | ·22721 872 | 1 181 | ·22257 027 | 1 115 | ·47976 138 | 2 495 | ·19920 846 | 5 834 |
| ·53 | ·22499 894 | 1 193 | ·22138 322 | 1 076 | ·48174 089 | 2 553 | ·19668 449 | 5 860 |
| ·54 | ·22279 109 | 1 203 | ·22018 541 | 1 040 | ·48369 487 | 2 611 | ·19410 192 | 5 885 |
| 0·55 | +0·22059 527 | +1 213 | −0·21897 720 | +1 001 | +0·48562 274 | −2 672 | −0·19146 050 | +5 909 |
| ·56 | ·21841 158 | 1 223 | ·21775 898 | 964 | ·48752 389 | 2 730 | ·18875 999 | 5 932 |
| ·57 | ·21624 012 | 1 233 | ·21653 112 | 929 | ·48939 774 | 2 790 | ·18600 016 | 5 955 |
| ·58 | ·21408 099 | 1 241 | ·21529 397 | 892 | ·49124 369 | 2 849 | ·18318 078 | 5 974 |
| ·59 | ·21193 427 | 1 251 | ·21404 790 | 857 | ·49306 115 | 2 908 | ·18030 166 | 5 994 |
| 0·60 | +0·20980 006 | +1 259 | −0·21279 326 | + 821 | +0·49484 953 | −2 970 | −0·17736 260 | +6 013 |
| ·61 | ·20767 844 | 1 266 | ·21153 041 | 786 | ·49660 821 | 3 030 | ·17436 341 | 6 030 |
| ·62 | ·20556 948 | 1 275 | ·21025 970 | 753 | ·49833 659 | 3 089 | ·17130 392 | 6 044 |
| ·63 | ·20347 327 | 1 281 | ·20898 146 | 717 | ·50003 408 | 3 150 | ·16818 399 | 6 061 |
| ·64 | ·20138 987 | 1 290 | ·20769 605 | 686 | ·50170 007 | 3 211 | ·16500 345 | 6 073 |
| 0·65 | +0·19931 937 | +1 295 | −0·20640 378 | + 651 | +0·50333 395 | −3 272 | −0·16176 218 | +6 084 |
| ·66 | ·19726 182 | 1 302 | ·20510 500 | 618 | ·50493 511 | 3 332 | ·15846 007 | 6 095 |
| ·67 | ·19521 729 | 1 308 | ·20380 004 | 588 | ·50650 295 | 3 394 | ·15509 701 | 6 105 |
| ·68 | ·19318 584 | 1 313 | ·20248 920 | 555 | ·50803 685 | 3 455 | ·15167 290 | 6 111 |
| ·69 | ·19116 752 | 1 320 | ·20117 281 | 523 | ·50953 620 | 3 515 | ·14818 768 | 6 117 |
| 0·70 | +0·18916 240 | +1 324 | −0·19985 119 | + 493 | +0·51100 040 | −3 578 | −0·14464 129 | +6 124 |
| ·71 | ·18717 052 | 1 328 | ·19852 464 | 462 | ·51242 882 | 3 637 | ·14103 366 | 6 124 |
| ·72 | ·18519 192 | 1 334 | ·19719 347 | 432 | ·51382 087 | 3 701 | ·13736 479 | 6 128 |
| ·73 | ·18322 666 | 1 338 | ·19585 798 | 403 | ·51517 591 | 3 759 | ·13363 464 | 6 127 |
| ·74 | ·18127 478 | 1 341 | ·19451 846 | 373 | ·51649 336 | 3 823 | ·12984 322 | 6 125 |
| 0·75 | +0·17933 631 | +1 344 | −0·19317 521 | + 345 | +0·51777 258 | −3 884 | −0·12599 055 | +6 123 |
| ·76 | ·17741 128 | 1 350 | ·19182 851 | 316 | ·51901 296 | 3 944 | ·12207 665 | 6 118 |
| ·77 | ·17549 975 | 1 350 | ·19047 865 | 288 | ·52021 390 | 4 005 | ·11810 157 | 6 111 |
| ·78 | ·17360 172 | 1 355 | ·18912 591 | 262 | ·52137 479 | 4 067 | ·11406 538 | 6 104 |
| ·79 | ·17171 724 | 1 356 | ·18777 055 | 233 | ·52249 501 | 4 128 | ·10996 815 | 6 093 |
| 0·80 | +0·16984 632 | +1 359 | −0·18641 286 | + 207 | +0·52357 395 | −4 188 | −0·10580 999 | +6 082 |
| ·81 | ·16798 899 | 1 360 | ·18505 310 | 181 | ·52461 101 | 4 250 | ·10159 101 | 6 069 |
| ·82 | ·16614 526 | 1 363 | ·18369 153 | 156 | ·52560 557 | 4 310 | ·09731 134 | 6 054 |
| ·83 | ·16431 516 | 1 364 | ·18232 840 | 129 | ·52655 703 | 4 370 | ·09297 113 | 6 037 |
| ·84 | ·16249 870 | 1 364 | ·18096 398 | 105 | ·52746 479 | 4 431 | ·08857 055 | 6 018 |
| 0·85 | +0·16069 588 | +1 367 | −0·17959 851 | + 81 | +0·52832 824 | −4 491 | −0·08410 979 | +5 999 |
| ·86 | ·15890 673 | 1 366 | ·17823 223 | 56 | ·52914 678 | 4 550 | ·07958 904 | 5 975 |
| ·87 | ·15713 124 | 1 367 | ·17686 539 | 32 | ·52991 982 | 4 610 | ·07500 854 | 5 952 |
| ·88 | ·15536 942 | 1 368 | ·17549 823 | + 10 | ·53064 676 | 4 670 | ·07036 852 | 5 925 |
| ·89 | ·15362 128 | 1 366 | ·17413 097 | − 13 | ·53132 700 | 4 729 | ·06566 925 | 5 898 |
| 0·90 | +0·15188 680 | +1 368 | −0·17276 384 | − 37 | +0·53195 995 | −4 788 | −0·06091 100 | +5 868 |
| ·91 | ·15016 600 | 1 366 | ·17139 708 | 58 | ·53254 502 | 4 846 | ·05609 407 | 5 835 |
| ·92 | ·14845 886 | 1 366 | ·17003 090 | 79 | ·53308 163 | 4 904 | ·05121 879 | 5 802 |
| ·93 | ·14676 538 | 1 365 | ·16866 551 | 101 | ·53356 920 | 4 962 | ·04628 549 | 5 767 |
| ·94 | ·14508 555 | 1 363 | ·16730 113 | 122 | ·53400 715 | 5 020 | ·04129 452 | 5 727 |
| 0·95 | +0·14341 935 | +1 363 | −0·16593 797 | − 142 | +0·53439 490 | −5 076 | −0·03624 628 | +5 688 |
| ·96 | ·14176 678 | 1 361 | ·16457 623 | 162 | ·53473 189 | 5 134 | ·03114 116 | 5 647 |
| ·97 | ·14012 782 | 1 359 | ·16321 611 | 182 | ·53501 754 | 5 190 | ·02597 957 | 5 601 |
| ·98 | ·13850 245 | 1 358 | ·16185 781 | 202 | ·53525 129 | 5 245 | ·02076 197 | 5 557 |
| ·99 | ·13689 066 | 1 355 | ·16050 153 | 219 | ·53543 259 | 5 301 | ·01548 880 | 5 506 |
| 1·00 | +0·13529 242 | +1 352 | −0·15914 744 | − 239 | +0·53556 088 | −5 355 | −0·01016 057 | +5 458 |

TABLE I—Ai$(x)$ AND Ai$'(x)$

| $x$ | Ai$(x)$ | $\delta^2$ | Ai$'(x)$ | $\delta^2$ | Ai$(-x)$ | $\delta^2$ | Ai$'(-x)$ | $\delta^2$ |
|---|---|---|---|---|---|---|---|---|
| 1·00 | +0·13529 242 | + 1 352 | −0·15914 744 | − 239 | +0·53556 088 | − 5 355 | −0·01016 057 | + 5 458 |
| ·01 | ·13370 770 | 1 351 | ·15779 574 | 257 | ·53563 562 | 5 410 | − ·00477 776 | 5 404 |
| ·02 | ·13213 649 | 1 348 | ·15644 661 | 273 | ·53565 626 | 5 464 | + ·00065 909 | 5 350 |
| ·03 | ·13057 876 | 1 345 | ·15510 021 | 293 | ·53562 226 | 5 517 | ·00614 944 | 5 293 |
| ·04 | ·12903 448 | 1 342 | ·15375 674 | 308 | ·53553 309 | 5 569 | ·01169 272 | 5 233 |
| 1·05 | +0·12750 362 | + 1 338 | −0·15241 635 | − 326 | +0·53538 823 | − 5 622 | +0·01728 833 | + 5 172 |
| ·06 | ·12598 614 | 1 336 | ·15107 922 | 341 | ·53518 715 | 5 672 | ·02293 566 | 5 109 |
| ·07 | ·12448 202 | 1 332 | ·14974 550 | 358 | ·53492 935 | 5 725 | ·02863 408 | 5 043 |
| ·08 | ·12299 122 | 1 328 | ·14841 536 | 372 | ·53461 430 | 5 773 | ·03438 293 | 4 974 |
| ·09 | ·12151 370 | 1 325 | ·14708 894 | 389 | ·53424 152 | 5 823 | ·04018 152 | 4 904 |
| 1·10 | +0·12004 943 | + 1 320 | −0·14576 641 | − 402 | +0·53381 051 | − 5 872 | +0·04602 915 | + 4 833 |
| ·11 | ·11859 836 | 1 317 | ·14444 790 | 418 | ·53332 078 | 5 920 | ·05192 511 | 4 755 |
| ·12 | ·11716 046 | 1 311 | ·14313 357 | 431 | ·53277 185 | 5 967 | ·05786 862 | 4 680 |
| ·13 | ·11573 567 | 1 309 | ·14182 355 | 446 | ·53216 325 | 6 013 | ·06385 893 | 4 600 |
| ·14 | ·11432 397 | 1 303 | ·14051 799 | 458 | ·53149 452 | 6 059 | ·06989 524 | 4 518 |
| 1·15 | +0·11292 530 | + 1 298 | −0·13921 701 | − 472 | +0·53076 520 | − 6 104 | +0·07597 673 | + 4 434 |
| ·16 | ·11153 961 | 1 295 | ·13792 075 | 485 | ·52997 484 | 6 148 | ·08210 256 | 4 347 |
| ·17 | ·11016 687 | 1 288 | ·13662 934 | 496 | ·52912 300 | 6 190 | ·08827 186 | 4 258 |
| ·18 | ·10880 701 | 1 284 | ·13534 289 | 509 | ·52820 926 | 6 233 | ·09448 374 | 4 167 |
| ·19 | ·10745 999 | 1 279 | ·13406 153 | 521 | ·52723 319 | 6 275 | ·10073 729 | 4 073 |
| 1·20 | +0·10612 576 | + 1 274 | −0·13278 538 | − 532 | +0·52619 437 | − 6 313 | +0·10703 157 | + 3 978 |
| ·21 | ·10480 427 | 1 267 | ·13151 455 | 543 | ·52509 242 | 6 354 | ·11336 563 | 3 878 |
| ·22 | ·10349 545 | 1 264 | ·13024 915 | 555 | ·52392 693 | 6 391 | ·11973 847 | 3 779 |
| ·23 | ·10219 927 | 1 256 | ·12898 930 | 563 | ·52269 753 | 6 430 | ·12614 910 | 3 676 |
| ·24 | ·10091 565 | 1 251 | ·12773 508 | 576 | ·52140 383 | 6 465 | ·13259 649 | 3 568 |
| 1·25 | +0·09964 454 | + 1 247 | −0·12648 662 | − 584 | +0·52004 548 | − 6 501 | +0·13907 956 | + 3 463 |
| ·26 | ·09838 590 | 1 238 | ·12524 400 | 595 | ·51862 212 | 6 534 | ·14559 726 | 3 351 |
| ·27 | ·09713 964 | 1 235 | ·12400 733 | 603 | ·51713 342 | 6 568 | ·15214 847 | 3 239 |
| ·28 | ·09590 573 | 1 227 | ·12277 669 | 612 | ·51557 904 | 6 598 | ·15873 207 | 3 124 |
| ·29 | ·09468 409 | 1 222 | ·12155 217 | 622 | ·51395 868 | 6 631 | ·16534 691 | 3 006 |
| 1·30 | +0·09347 467 | + 1 214 | −0·12033 387 | − 629 | +0·51227 201 | − 6 660 | +0·17199 181 | + 2 886 |
| ·31 | ·09227 739 | 1 210 | ·11912 186 | 637 | ·51051 874 | 6 687 | ·17866 557 | 2 766 |
| ·32 | ·09109 221 | 1 202 | ·11791 622 | 647 | ·50869 860 | 6 715 | ·18536 699 | 2 639 |
| ·33 | ·08991 905 | 1 195 | ·11671 705 | 652 | ·50681 131 | 6 740 | ·19209 480 | 2 513 |
| ·34 | ·08875 784 | 1 191 | ·11552 440 | 661 | ·50485 662 | 6 765 | ·19884 774 | 2 384 |
| 1·35 | +0·08760 854 | + 1 182 | −0·11433 836 | − 668 | +0·50283 428 | − 6 788 | +0·20562 452 | + 2 252 |
| ·36 | ·08647 106 | 1 175 | ·11315 900 | 673 | ·50074 406 | 6 811 | ·21242 382 | 2 119 |
| ·37 | ·08534 533 | 1 171 | ·11198 637 | 681 | ·49858 573 | 6 830 | ·21924 431 | 1 981 |
| ·38 | ·08423 131 | 1 161 | ·11082 055 | 688 | ·49635 910 | 6 849 | ·22608 461 | 1 844 |
| ·39 | ·08312 890 | 1 156 | ·10966 161 | 692 | ·49406 398 | 6 868 | ·23294 335 | 1 703 |
| 1·40 | +0·08203 805 | + 1 148 | −0·10850 959 | − 699 | +0·49170 018 | − 6 883 | +0·23981 912 | + 1 559 |
| ·41 | ·08095 868 | 1 143 | ·10736 456 | 704 | ·48926 755 | 6 900 | ·24671 048 | 1 414 |
| ·42 | ·07989 074 | 1 133 | ·10622 657 | 710 | ·48676 592 | 6 911 | ·25361 598 | 1 265 |
| ·43 | ·07883 413 | 1 128 | ·10509 568 | 714 | ·48419 518 | 6 924 | ·26053 413 | 1 117 |
| ·44 | ·07778 880 | 1 120 | ·10397 193 | 720 | ·48155 520 | 6 934 | ·26746 345 | 964 |
| 1·45 | +0·07675 467 | + 1 113 | −0·10285 538 | − 723 | +0·47884 588 | − 6 943 | +0·27440 241 | + 810 |
| ·46 | ·07573 167 | 1 105 | ·10174 606 | 729 | ·47606 713 | 6 951 | ·28134 947 | 652 |
| ·47 | ·07471 972 | 1 099 | ·10064 403 | 732 | ·47321 887 | 6 956 | ·28830 305 | 494 |
| ·48 | ·07371 876 | 1 091 | ·09954 932 | 736 | ·47030 105 | 6 960 | ·29526 157 | 333 |
| ·49 | ·07272 871 | 1 084 | ·09846 197 | 739 | ·46731 363 | 6 963 | ·30222 342 | 170 |
| 1·50 | +0·07174 950 | + 1 076 | −0·09738 201 | − 744 | +0·46425 658 | − 6 964 | +0·30918 697 | + 4 |

TABLE I—Ai(x) AND Ai'(x)

| x | Ai(x) | $\delta^2$ | Ai'(x) | $\delta^2$ | Ai(−x) | $\delta^2$ | Ai'(−x) | $\delta^2$ |
|---|---|---|---|---|---|---|---|---|
| 1·50 | +0·07174 950 | +1 076 | −0·09738 201 | −744 | +0·46425 658 | −6 964 | +0·30918 697 | + 4 |
| ·51 | ·07078 105 | 1 068 | ·09630 949 | 747 | ·46112 989 | 6 963 | ·31615 056 | − 162 |
| ·52 | ·06982 328 | 1 062 | ·09524 444 | 748 | ·45793 357 | 6 960 | ·32311 253 | 333 |
| ·53 | ·06887 613 | 1 054 | ·09418 687 | 754 | ·45466 765 | 6 956 | ·33007 117 | 503 |
| ·54 | ·06793 952 | 1 046 | ·09313 684 | 754 | ·45133 217 | 6 951 | ·33702 478 | 677 |
| 1·55 | +0·06701 337 | +1 039 | −0·09209 435 | −757 | +0·44792 718 | −6 943 | +0·34397 162 | − 852 |
| ·56 | ·06609 761 | 1 031 | ·09105 943 | 760 | ·44445 276 | 6 933 | ·35090 994 | 1 030 |
| ·57 | ·06519 216 | 1 023 | ·09003 211 | 761 | ·44090 901 | 6 922 | ·35783 796 | 1 210 |
| ·58 | ·06429 694 | 1 016 | ·08901 240 | 764 | ·43729 604 | 6 909 | ·36475 388 | 1 390 |
| ·59 | ·06341 188 | 1 009 | ·08800 033 | 765 | ·43361 398 | 6 894 | ·37165 590 | 1 573 |
| 1·60 | +0·06253 691 | +1 000 | −0·08699 591 | −766 | +0·42986 298 | −6 878 | +0·37854 219 | − 1 758 |
| ·61 | ·06167 194 | 993 | ·08599 915 | 769 | ·42604 320 | 6 860 | ·38541 090 | 1 945 |
| ·62 | ·06081 690 | 985 | ·08501 008 | 768 | ·42215 482 | 6 838 | ·39226 016 | 2 134 |
| ·63 | ·05997 171 | 978 | ·08402 869 | 770 | ·41819 806 | 6 816 | ·39908 808 | 2 322 |
| ·64 | ·05913 630 | 970 | ·08305 500 | 771 | ·41417 314 | 6 793 | ·40589 278 | 2 516 |
| 1·65 | +0·05831 059 | + 961 | −0·08208 902 | −771 | +0·41008 029 | −6 766 | +0·41267 232 | − 2 708 |
| ·66 | ·05749 449 | 956 | ·08113 075 | 772 | ·40591 978 | 6 738 | ·41942 478 | 2 903 |
| ·67 | ·05668 795 | 945 | ·08018 020 | 773 | ·40169 189 | 6 708 | ·42614 821 | 3 101 |
| ·68 | ·05589 086 | 940 | ·07923 738 | 771 | ·39739 692 | 6 676 | ·43284 063 | 3 297 |
| ·69 | ·05510 317 | 931 | ·07830 227 | 773 | ·39303 519 | 6 642 | ·43950 008 | 3 498 |
| 1·70 | +0·05432 479 | + 924 | −0·07737 489 | −772 | +0·38860 704 | −6 607 | +0·44612 455 | − 3 697 |
| ·71 | ·05355 565 | 915 | ·07645 523 | 772 | ·38411 282 | 6 567 | ·45271 205 | 3 901 |
| ·72 | ·05279 566 | 909 | ·07554 329 | 771 | ·37955 293 | 6 529 | ·45926 054 | 4 104 |
| ·73 | ·05204 476 | 900 | ·07463 906 | 770 | ·37492 775 | 6 486 | ·46576 799 | 4 309 |
| ·74 | ·05130 286 | 892 | ·07374 253 | 771 | ·37023 771 | 6 442 | ·47223 235 | 4 514 |
| 1·75 | +0·05056 988 | + 886 | −0·07285 371 | −770 | +0·36548 325 | −6 395 | +0·47865 157 | − 4 722 |
| ·76 | ·04984 576 | 876 | ·07197 259 | 767 | ·36066 484 | 6 348 | ·48502 357 | 4 929 |
| ·77 | ·04913 040 | 871 | ·07109 914 | 768 | ·35578 295 | 6 298 | ·49134 628 | 5 140 |
| ·78 | ·04842 375 | 861 | ·07023 337 | 765 | ·35083 808 | 6 244 | ·49761 759 | 5 349 |
| ·79 | ·04772 571 | 855 | ·06937 525 | 765 | ·34583 077 | 6 190 | ·50383 541 | 5 560 |
| 1·80 | +0·04703 622 | + 846 | −0·06852 478 | −763 | +0·34076 156 | −6 134 | +0·50999 763 | − 5 773 |
| ·81 | ·04635 519 | 839 | ·06768 194 | 762 | ·33563 101 | 6 074 | ·51610 212 | 5 985 |
| ·82 | ·04568 255 | 832 | ·06684 672 | 759 | ·33043 972 | 6 015 | ·52214 676 | 6 199 |
| ·83 | ·04501 823 | 824 | ·06601 909 | 758 | ·32518 828 | 5 950 | ·52812 941 | 6 412 |
| ·84 | ·04436 215 | 815 | ·06519 904 | 757 | ·31987 734 | 5 885 | ·53404 794 | 6 629 |
| 1·85 | +0·04371 422 | + 810 | −0·06438 656 | −753 | +0·31450 755 | −5 819 | +0·53990 018 | − 6 842 |
| ·86 | ·04307 439 | 801 | ·06358 161 | 752 | ·30907 957 | 5 749 | ·54568 400 | 7 060 |
| ·87 | ·04244 257 | 793 | ·06278 418 | 750 | ·30359 410 | 5 677 | ·55139 722 | 7 275 |
| ·88 | ·04181 868 | 787 | ·06199 425 | 747 | ·29805 186 | 5 602 | ·55703 769 | 7 491 |
| ·89 | ·04120 266 | 778 | ·06121 179 | 745 | ·29245 360 | 5 528 | ·56260 325 | 7 709 |
| 1·90 | +0·04059 442 | + 772 | −0·06043 678 | −743 | +0·28680 006 | −5 449 | +0·56809 172 | − 7 926 |
| ·91 | ·03999 390 | 763 | ·05966 920 | 739 | ·28109 203 | 5 369 | ·57350 093 | 8 143 |
| ·92 | ·03940 101 | 757 | ·05890 901 | 737 | ·27533 031 | 5 286 | ·57882 871 | 8 360 |
| ·93 | ·03881 569 | 749 | ·05815 619 | 734 | ·26951 573 | 5 201 | ·58407 289 | 8 578 |
| ·94 | ·03823 786 | 742 | ·05741 071 | 732 | ·26364 914 | 5 115 | ·58923 129 | 8 794 |
| 1·95 | +0·03766 745 | + 735 | −0·05667 255 | −728 | +0·25773 140 | −5 026 | +0·59430 175 | − 9 012 |
| ·96 | ·03710 439 | 727 | ·05594 167 | 726 | ·25176 340 | 4 934 | ·59928 209 | 9 228 |
| ·97 | ·03654 860 | 719 | ·05521 805 | 722 | ·24574 606 | 4 841 | ·60417 015 | 9 444 |
| ·98 | ·03600 000 | 714 | ·05450 165 | 719 | ·23968 031 | 4 745 | ·60896 377 | 9 662 |
| ·99 | ·03545 854 | 705 | ·05379 244 | 715 | ·23356 711 | 4 648 | ·61366 077 | 9 875 |
| 2·00 | +0·03492 413 | + 699 | −0·05309 038 | −714 | +0·22740 743 | −4 548 | +0·61825 902 | −10 091 |

# TABLE I—Ai($x$) AND Ai$'(x)$

| $x$ | Ai($-x$) | $\delta^2$ | Ai$'(-x)$ | $\delta^2$ | $x$ | Ai($-x$) | $\delta^2$ | Ai$'(-x)$ | $\delta^2$ |
|---|---|---|---|---|---|---|---|---|---|
| 2·00 | +0·22740 743 | − 4 548 | +0·61825 902 | − 10 091 | 2·50 | −0·11232 507 | + 2 809 | +0·67885 273 | − 18 093 |
| ·01 | ·22120 227 | 4 447 | ·62275 636 | 10 306 | ·51 | ·11909 925 | 2 989 | ·67595 403 | 18 158 |
| ·02 | ·21495 264 | 4 341 | ·62715 064 | 10 519 | ·52 | ·12584 354 | 3 171 | ·67287 375 | 18 214 |
| ·03 | ·20865 960 | 4 235 | ·63143 973 | 10 731 | ·53 | ·13255 612 | 3 354 | ·66961 133 | 18 266 |
| ·04 | ·20232 421 | 4 128 | ·63562 151 | 10 943 | ·54 | ·13923 516 | 3 537 | ·66616 625 | 18 313 |
| 2·05 | +0·19594 754 | − 4 017 | +0·63969 386 | − 11 155 | 2·55 | −0·14587 883 | + 3 719 | +0·66253 804 | − 18 353 |
| ·06 | ·18953 070 | 3 904 | ·64365 466 | 11 364 | ·56 | ·15248 531 | 3 904 | ·65872 630 | 18 387 |
| ·07 | ·18307 482 | 3 789 | ·64750 182 | 11 572 | ·57 | ·15905 275 | 4 088 | ·65473 069 | 18 418 |
| ·08 | ·17658 105 | 3 673 | ·65123 326 | 11 779 | ·58 | ·16557 931 | 4 272 | ·65055 090 | 18 438 |
| ·09 | ·17005 055 | 3 554 | ·65484 691 | 11 987 | ·59 | ·17206 315 | 4 456 | ·64618 673 | 18 457 |
| 2·10 | +0·16348 451 | − 3 433 | +0·65834 069 | − 12 189 | 2·60 | −0·17850 243 | + 4 641 | +0·64163 799 | − 18 467 |
| ·11 | ·15688 414 | 3 310 | ·66171 258 | 12 394 | ·61 | ·18489 530 | 4 827 | ·63690 458 | 18 472 |
| ·12 | ·15025 067 | 3 185 | ·66496 053 | 12 594 | ·62 | ·19123 990 | 5 009 | ·63198 645 | 18 470 |
| ·13 | ·14358 535 | 3 058 | ·66808 254 | 12 795 | ·63 | ·19753 441 | 5 196 | ·62688 362 | 18 461 |
| ·14 | ·13688 945 | 2 930 | ·67107 660 | 12 991 | ·64 | ·20377 696 | 5 379 | ·62159 618 | 18 448 |
| 2·15 | +0·13016 425 | − 2 798 | +0·67394 075 | − 13 188 | 2·65 | −0·20996 572 | + 5 565 | +0·61612 426 | − 18 427 |
| ·16 | ·12341 107 | 2 665 | ·67667 302 | 13 383 | ·66 | ·21609 883 | 5 747 | ·61046 807 | 18 398 |
| ·17 | ·11663 124 | 2 531 | ·67927 146 | 13 572 | ·67 | ·22217 447 | 5 933 | ·60462 790 | 18 365 |
| ·18 | ·10982 610 | 2 394 | ·68173 418 | 13 765 | ·68 | ·22819 078 | 6 115 | ·59860 408 | 18 324 |
| ·19 | ·10299 702 | 2 256 | ·68405 925 | 13 950 | ·69 | ·23414 594 | 6 299 | ·59239 702 | 18 276 |
| 2·20 | +0·09614 538 | − 2 115 | +0·68624 482 | − 14 135 | 2·70 | −0·24003 811 | + 6 481 | +0·58600 720 | − 18 222 |
| ·21 | ·08927 259 | 1 973 | ·68828 904 | 14 319 | ·71 | ·24586 547 | 6 662 | ·57943 516 | 18 161 |
| ·22 | ·08238 007 | 1 828 | ·69019 007 | 14 498 | ·72 | ·25162 621 | 6 845 | ·57268 151 | 18 093 |
| ·23 | ·07546 927 | 1 683 | ·69194 612 | 14 676 | ·73 | ·25731 850 | 7 024 | ·56574 693 | 18 017 |
| ·24 | ·06854 164 | 1 535 | ·69355 541 | 14 849 | ·74 | ·26294 055 | 7 205 | ·55863 218 | 17 936 |
| 2·25 | +0·06159 866 | − 1 386 | +0·69501 621 | − 15 023 | 2·75 | −0·26849 055 | + 7 384 | +0·55133 807 | − 17 845 |
| ·26 | ·05464 182 | 1 235 | ·69632 678 | 15 189 | ·76 | ·27396 671 | 7 561 | ·54386 551 | 17 751 |
| ·27 | ·04767 263 | 1 082 | ·69748 546 | 15 356 | ·77 | ·27936 726 | 7 738 | ·53621 544 | 17 645 |
| ·28 | ·04069 262 | 927 | ·69849 058 | 15 519 | ·78 | ·28469 043 | 7 914 | ·52838 892 | 17 536 |
| ·29 | ·03370 334 | 773 | ·69934 051 | 15 678 | ·79 | ·28993 446 | 8 090 | ·52038 704 | 17 418 |
| 2·30 | +0·02670 633 | − 613 | +0·70003 366 | − 15 833 | 2·80 | −0·29509 759 | + 8 262 | +0·51221 098 | − 17 292 |
| ·31 | ·01970 319 | 455 | ·70056 848 | 15 985 | ·81 | ·30017 810 | 8 435 | ·50386 200 | 17 159 |
| ·32 | ·01269 550 | 295 | ·70094 345 | 16 135 | ·82 | ·30517 426 | 8 606 | ·49534 143 | 17 021 |
| ·33 | + ·00568 486 | − 132 | ·70115 707 | 16 281 | ·83 | ·31008 436 | 8 775 | ·48665 065 | 16 872 |
| ·34 | − ·00132 710 | + 31 | ·70120 788 | 16 420 | ·84 | ·31490 671 | 8 943 | ·47779 115 | 16 717 |
| 2·35 | −0·00833 875 | + 196 | +0·70109 449 | − 16 559 | 2·85 | −0·31963 963 | + 9 110 | +0·46876 448 | − 16 556 |
| ·36 | ·01534 844 | 362 | ·70081 551 | 16 692 | ·86 | ·32428 145 | 9 274 | ·45957 225 | 16 386 |
| ·37 | ·02235 451 | 531 | ·70036 961 | 16 823 | ·87 | ·32883 053 | 9 438 | ·45021 616 | 16 209 |
| ·38 | ·02935 527 | 698 | ·69975 548 | 16 947 | ·88 | ·33328 523 | 9 598 | ·44069 798 | 16 024 |
| ·39 | ·03634 905 | 869 | ·69897 188 | 17 068 | ·89 | ·33764 395 | 9 757 | ·43101 956 | 15 833 |
| 2·40 | −0·04333 414 | + 1 040 | +0·69801 760 | − 17 186 | 2·90 | −0·34190 510 | + 9 916 | +0·42118 281 | − 15 632 |
| ·41 | ·05030 883 | 1 213 | ·69689 146 | 17 297 | ·91 | ·34606 709 | 10 070 | ·41118 974 | 15 426 |
| ·42 | ·05727 139 | 1 385 | ·69559 235 | 17 406 | ·92 | ·35012 838 | 10 224 | ·40104 241 | 15 211 |
| ·43 | ·06422 010 | 1 562 | ·69411 918 | 17 509 | ·93 | ·35408 743 | 10 374 | ·39074 297 | 14 989 |
| ·44 | ·07115 319 | 1 735 | ·69247 092 | 17 608 | ·94 | ·35794 274 | 10 524 | ·38029 364 | 14 760 |
| 2·45 | −0·07806 893 | + 1 914 | +0·69064 658 | − 17 701 | 2·95 | −0·36169 281 | + 10 669 | +0·36969 671 | − 14 521 |
| ·46 | ·08496 553 | 2 089 | ·68864 523 | 17 789 | ·96 | ·36533 619 | 10 814 | ·35895 457 | 14 279 |
| ·47 | ·09184 124 | 2 269 | ·68646 599 | 17 874 | ·97 | ·36887 143 | 10 956 | ·34806 964 | 14 025 |
| ·48 | ·09869 426 | 2 448 | ·68410 801 | 17 953 | ·98 | ·37229 711 | 11 093 | ·33704 446 | 13 767 |
| ·49 | ·10552 280 | 2 627 | ·68157 050 | 18 026 | ·99 | ·37561 186 | 11 232 | ·32588 161 | 13 499 |
| 2·50 | −0·11232 507 | + 2 809 | +0·67885 273 | − 18 093 | 3·00 | −0·37881 429 | + 11 363 | +0·31458 377 | − 13 225 |

## TABLE I—Ai($x$) AND Ai$'$($x$)

| $x$ | Ai($-x$) | $\delta^2$ | Ai$'$($-x$) | $\delta^2$ | $x$ | Ai($-x$) | $\delta^2$ | Ai$'$($-x$) | $\delta^2$ |
|---|---|---|---|---|---|---|---|---|---|
| 3·00 | −0·37881 429 | +11 363 | +0·31458 377 | − 13 225 | 3·50 | −0·37553 382 | +13 143 | −0·34344 343 | + 8 265 |
| ·01 | ·38190 309 | 11 495 | ·30315 368 | 12 944 | ·51 | ·37203 381 | 13 058 | ·35654 491 | 8 795 |
| ·02 | ·38487 694 | 11 624 | ·29159 415 | 12 653 | ·52 | ·36840 322 | 12 968 | ·36955 844 | 9 324 |
| ·03 | ·38773 455 | 11 747 | ·27990 809 | 12 359 | ·53 | ·36464 295 | 12 871 | ·38247 873 | 9 856 |
| ·04 | ·39047 469 | 11 871 | ·26809 844 | 12 054 | ·54 | ·36075 397 | 12 770 | ·39530 046 | 10 385 |
| 3·05 | −0·39309 612 | +11 989 | +0·25616 825 | − 11 743 | 3·55 | −0·35673 729 | +12 664 | −0·40801 834 | +10 918 |
| ·06 | ·39559 766 | 12 105 | ·24412 063 | 11 426 | ·56 | ·35259 397 | 12 552 | ·42062 704 | 11 448 |
| ·07 | ·39797 815 | 12 218 | ·23195 875 | 11 101 | ·57 | ·34832 513 | 12 435 | ·43312 126 | 11 979 |
| ·08 | ·40023 646 | 12 326 | ·21968 586 | 10 767 | ·58 | ·34393 194 | 12 312 | ·44549 569 | 12 510 |
| ·09 | ·40237 151 | 12 434 | ·20730 530 | 10 429 | ·59 | ·33941 563 | 12 184 | ·45774 502 | 13 038 |
| 3·10 | −0·40438 222 | +12 535 | +0·19482 045 | − 10 083 | 3·60 | −0·33477 748 | +12 052 | −0·46986 397 | +13 568 |
| ·11 | ·40626 758 | 12 635 | ·18223 477 | 9 730 | ·61 | ·33001 881 | 11 914 | ·48184 724 | 14 094 |
| ·12 | ·40802 659 | 12 730 | ·16955 179 | 9 369 | ·62 | ·32514 100 | 11 769 | ·49368 957 | 14 620 |
| ·13 | ·40965 830 | 12 821 | ·15677 512 | 9 004 | ·63 | ·32014 550 | 11 621 | ·50538 570 | 15 144 |
| ·14 | ·41116 180 | 12 911 | ·14390 841 | 8 629 | ·64 | ·31503 379 | 11 466 | ·51693 039 | 15 666 |
| 3·15 | −0·41253 619 | +12 995 | +0·13095 541 | − 8 250 | 3·65 | −0·30980 742 | +11 309 | −0·52831 842 | +16 185 |
| ·16 | ·41378 063 | 13 075 | ·11791 991 | 7 863 | ·66 | ·30446 796 | 11 142 | ·53954 460 | 16 702 |
| ·17 | ·41489 432 | 13 151 | ·10480 578 | 7 472 | ·67 | ·29901 708 | 10 974 | ·55060 376 | 17 217 |
| ·18 | ·41587 650 | 13 226 | ·09161 693 | 7 071 | ·68 | ·29345 646 | 10 798 | ·56149 075 | 17 728 |
| ·19 | ·41672 642 | 13 292 | ·07835 737 | 6 666 | ·69 | ·28778 786 | 10 620 | ·57220 046 | 18 237 |
| 3·20 | −0·41744 342 | +13 358 | +0·06503 115 | − 6 256 | 3·70 | −0·28201 306 | +10 433 | −0·58272 780 | +18 739 |
| ·21 | ·41802 684 | 13 419 | ·05164 237 | 5 836 | ·71 | ·27613 393 | 10 245 | ·59306 775 | 19 242 |
| ·22 | ·41847 607 | 13 474 | ·03819 523 | 5 415 | ·72 | ·27015 235 | 10 049 | ·60321 528 | 19 738 |
| ·23 | ·41879 056 | 13 526 | ·02469 394 | 4 986 | ·73 | ·26407 028 | 9 850 | ·61316 543 | 20 229 |
| ·24 | ·41896 979 | 13 575 | + ·01114 279 | 4 549 | ·74 | ·25788 971 | 9 644 | ·62291 329 | 20 718 |
| 3·25 | −0·41901 327 | +13 618 | −0·00245 385 | − 4 110 | 3·75 | −0·25161 270 | + 9 435 | −0·63245 397 | +21 201 |
| ·26 | ·41892 057 | 13 656 | ·01609 159 | 3 664 | ·76 | ·24524 134 | 9 221 | ·64178 264 | 21 678 |
| ·27 | ·41869 131 | 13 690 | ·02976 597 | 3 214 | ·77 | ·23877 777 | 9 001 | ·65089 453 | 22 150 |
| ·28 | ·41832 515 | 13 722 | ·04347 249 | 2 757 | ·78 | ·23222 419 | 8 778 | ·65978 492 | 22 618 |
| ·29 | ·41782 177 | 13 745 | ·05720 658 | 2 295 | ·79 | ·22558 283 | 8 549 | ·66844 913 | 23 077 |
| 3·30 | −0·41718 094 | +13 767 | −0·07096 362 | − 1 830 | 3·80 | −0·21885 598 | + 8 317 | −0·67688 257 | +23 533 |
| ·31 | ·41640 244 | 13 783 | ·08473 896 | 1 358 | ·81 | ·21204 596 | 8 078 | ·68508 068 | 23 980 |
| ·32 | ·41548 611 | 13 793 | ·09852 788 | 884 | ·82 | ·20515 516 | 7 836 | ·69303 899 | 24 422 |
| ·33 | ·41443 185 | 13 800 | ·11232 564 | − 404 | ·83 | ·19818 600 | 7 591 | ·70075 308 | 24 857 |
| ·34 | ·41323 959 | 13 803 | ·12612 744 | + 82 | ·84 | ·19114 093 | 7 339 | ·70821 860 | 25 283 |
| 3·35 | −0·41190 930 | +13 797 | −0·13992 842 | + 567 | 3·85 | −0·18402 247 | + 7 085 | −0·71543 129 | +25 703 |
| ·36 | ·41044 104 | 13 791 | ·15372 373 | 1 062 | ·86 | ·17683 316 | 6 825 | ·72238 695 | 26 116 |
| ·37 | ·40883 487 | 13 778 | ·16750 842 | 1 557 | ·87 | ·16957 560 | 6 563 | ·72908 145 | 26 519 |
| ·38 | ·40709 092 | 13 759 | ·18127 754 | 2 057 | ·88 | ·16225 241 | 6 294 | ·73551 076 | 26 914 |
| ·39 | ·40520 938 | 13 736 | ·19502 609 | 2 559 | ·89 | ·15486 628 | 6 024 | ·74167 093 | 27 301 |
| 3·40 | −0·40319 048 | +13 708 | −0·20874 905 | + 3 066 | 3·90 | −0·14741 991 | + 5 750 | −0·74755 809 | +27 681 |
| ·41 | ·40103 450 | 13 674 | ·22244 135 | 3 575 | ·91 | ·13991 604 | 5 470 | ·75316 844 | 28 049 |
| ·42 | ·39874 178 | 13 638 | ·23609 790 | 4 087 | ·92 | ·13235 747 | 5 188 | ·75849 830 | 28 408 |
| ·43 | ·39631 268 | 13 592 | ·24971 358 | 4 603 | ·93 | ·12474 702 | 4 902 | ·76354 408 | 28 759 |
| ·44 | ·39374 766 | 13 545 | ·26328 323 | 5 119 | ·94 | ·11708 755 | 4 613 | ·76830 227 | 29 100 |
| 3·45 | −0·39104 719 | +13 491 | −0·27680 169 | + 5 640 | 3·95 | −0·10938 195 | + 4 321 | −0·77276 946 | +29 429 |
| ·46 | ·38821 181 | 13 431 | ·29026 375 | 6 161 | ·96 | ·10163 314 | 4 024 | ·77694 236 | 29 750 |
| ·47 | ·38524 212 | 13 368 | ·30366 420 | 6 685 | ·97 | ·09384 409 | 3 725 | ·78081 776 | 30 059 |
| ·48 | ·38213 875 | 13 298 | ·31699 780 | 7 210 | ·98 | ·08601 779 | 3 424 | ·78439 257 | 30 358 |
| ·49 | ·37890 240 | 13 223 | ·33025 930 | 7 737 | ·99 | ·07815 725 | 3 118 | ·78766 380 | 30 645 |
| 3·50 | −0·37553 382 | +13 143 | −0·34344 343 | + 8 265 | 4·00 | −0·07026 553 | + 2 810 | −0·79062 858 | +30 922 |

## TABLE I—Ai($x$) AND Ai$'$($x$)

| $x$ | Ai($-x$) | $\delta^2$ | Ai$'$($-x$) | $\delta^2_m$ | $x$ | Ai($-x$) | $\delta^2$ | Ai$'$($-x$) | $\delta^2_m$ |
|---|---|---|---|---|---|---|---|---|---|
| 4·00 | − 0·07026 553 | + 2 810 | − 0·79062 858 | + 30 924 | 4·50 | + 0·29215 278 | − 13 147 | − 0·52336 253 | + 26 474 |
| ·01 | ·06234 571 | 2 500 | ·79328 414 | 31 188 | ·51 | ·29732 023 | 13 408 | ·51008 411 | 25 980 |
| ·02 | ·05440 089 | 2 187 | ·79562 784 | 31 441 | ·52 | ·30235 360 | 13 667 | ·49654 592 | 25 469 |
| ·03 | ·04643 420 | 1 871 | ·79765 715 | 31 684 | ·53 | ·30725 030 | 13 917 | ·48275 307 | 24 941 |
| ·04 | ·03844 880 | 1 553 | ·79936 965 | 31 910 | ·54 | ·31200 783 | 14 165 | ·46871 083 | 24 404 |
| 4·05 | − 0·03044 787 | + 1 233 | − 0·80076 307 | + 32 129 | 4·55 | + 0·31662 371 | − 14 407 | − 0·45442 459 | + 23 842 |
| ·06 | ·02243 461 | 910 | ·80183 523 | 32 331 | ·56 | ·32109 552 | 14 640 | ·43989 995 | 23 273 |
| ·07 | ·01441 225 | 588 | ·80258 410 | 32 523 | ·57 | ·32542 093 | 14 872 | ·42514 261 | 22 683 |
| ·08 | − ·00638 401 | + 259 | ·80300 777 | 32 699 | ·58 | ·32959 762 | 15 095 | ·41015 846 | 22 084 |
| ·09 | + ·00164 682 | − 67 | ·80310 447 | 32 866 | ·59 | ·33362 336 | 15 312 | ·39495 350 | 21 465 |
| 4·10 | + 0·00967 698 | − 397 | − 0·80287 254 | + 33 017 | 4·60 | + 0·33749 598 | − 15 525 | − 0·37953 391 | + 20 835 |
| ·11 | ·01770 317 | 728 | ·80231 047 | 33 152 | ·61 | ·34121 335 | 15 730 | ·36390 600 | 20 190 |
| ·12 | ·02572 208 | 1 059 | ·80141 690 | 33 278 | ·62 | ·34477 342 | 15 927 | ·34807 622 | 19 531 |
| ·13 | ·03373 040 | 1 394 | ·80019 058 | 33 385 | ·63 | ·34817 422 | 16 120 | ·33205 116 | 18 858 |
| ·14 | ·04172 478 | 1 727 | ·79863 043 | 33 483 | ·64 | ·35141 382 | 16 306 | ·31583 755 | 18 169 |
| 4·15 | + 0·04970 189 | − 2 063 | − 0·79673 548 | + 33 564 | 4·65 | + 0·35449 036 | − 16 482 | − 0·29944 227 | + 17 472 |
| ·16 | ·05765 837 | 2 398 | ·79450 492 | 33 628 | ·66 | ·35740 208 | 16 655 | ·28287 230 | 16 756 |
| ·17 | ·06559 087 | 2 736 | ·79193 810 | 33 681 | ·67 | ·36014 725 | 16 818 | ·26613 479 | 16 031 |
| ·18 | ·07349 601 | 3 071 | ·78903 450 | 33 719 | ·68 | ·36272 424 | 16 975 | ·24923 699 | 15 294 |
| ·19 | ·08137 044 | 3 411 | ·78579 374 | 33 740 | ·69 | ·36513 148 | 17 124 | ·23218 628 | 14 541 |
| 4·20 | + 0·08921 076 | − 3 746 | − 0·78221 561 | + 33 747 | 4·70 | + 0·36736 748 | − 17 265 | − 0·21499 018 | + 13 781 |
| ·21 | ·09701 362 | 4 084 | ·77830 004 | 33 738 | ·71 | ·36943 083 | 17 400 | ·19765 630 | 13 003 |
| ·22 | ·10477 564 | 4 422 | ·77404 712 | 33 715 | ·72 | ·37132 018 | 17 526 | ·18019 240 | 12 220 |
| ·23 | ·11249 344 | 4 758 | ·76945 708 | 33 673 | ·73 | ·37303 427 | 17 644 | ·16260 632 | 11 423 |
| ·24 | ·12016 366 | 5 095 | ·76453 033 | 33 620 | ·74 | ·37457 192 | 17 754 | ·14490 603 | 10 615 |
| 4·25 | + 0·12778 293 | − 5 431 | − 0·75926 741 | + 33 548 | 4·75 | + 0·37593 203 | − 17 855 | − 0·12709 961 | + 9 798 |
| ·26 | ·13534 789 | 5 766 | ·75366 904 | 33 461 | ·76 | ·37711 359 | 17 951 | ·10919 523 | 8 970 |
| ·27 | ·14285 519 | 6 099 | ·74773 609 | 33 359 | ·77 | ·37811 564 | 18 035 | ·09120 117 | 8 133 |
| ·28 | ·15030 150 | 6 433 | ·74146 958 | 33 240 | ·78 | ·37893 734 | 18 113 | ·07312 580 | 7 284 |
| ·29 | ·15768 348 | 6 765 | ·73487 070 | 33 104 | ·79 | ·37957 791 | 18 180 | ·05497 760 | 6 433 |
| 4·30 | + 0·16499 781 | − 7 095 | − 0·72794 081 | + 32 953 | 4·80 | + 0·38003 668 | − 18 242 | − 0·03676 510 | + 5 564 |
| ·31 | ·17224 119 | 7 423 | ·72068 142 | 32 785 | ·81 | ·38031 303 | 18 292 | ·01849 697 | 4 695 |
| ·32 | ·17941 034 | 7 750 | ·71309 421 | 32 602 | ·82 | ·38040 646 | 18 335 | − ·00018 191 | 3 813 |
| ·33 | ·18650 199 | 8 075 | ·70518 101 | 32 401 | ·83 | ·38031 654 | 18 368 | + ·01817 127 | 2 926 |
| ·34 | ·19351 289 | 8 399 | ·69694 383 | 32 184 | ·84 | ·38004 294 | 18 394 | ·03655 370 | 2 031 |
| 4·35 | + 0·20043 980 | − 8 719 | − 0·68838 484 | + 31 950 | 4·85 | + 0·37958 540 | − 18 409 | + 0·05495 643 | + 1 133 |
| ·36 | ·20727 952 | 9 037 | ·67950 638 | 31 703 | ·86 | ·37894 377 | 18 416 | ·07337 047 | + 222 |
| ·37 | ·21402 887 | 9 353 | ·67031 093 | 31 434 | ·87 | ·37811 798 | 18 413 | ·09178 673 | − 686 |
| ·38 | ·22068 469 | 9 666 | ·66080 117 | 31 152 | ·88 | ·37710 806 | 18 403 | ·11019 611 | 1 608 |
| ·39 | ·22724 385 | 9 975 | ·65097 992 | 30 852 | ·89 | ·37591 411 | 18 381 | ·12858 941 | 2 526 |
| 4·40 | + 0·23370 326 | − 10 283 | − 0·64085 018 | + 30 534 | 4·90 | + 0·37453 635 | − 18 351 | + 0·14695 743 | − 3 457 |
| ·41 | ·24005 984 | 10 587 | ·63041 512 | 30 206 | ·91 | ·37297 508 | 18 313 | ·16529 088 | 4 385 |
| ·42 | ·24631 055 | 10 886 | ·61967 804 | 29 854 | ·92 | ·37123 068 | 18 263 | ·18358 047 | 5 320 |
| ·43 | ·25245 240 | 11 183 | ·60864 245 | 29 489 | ·93 | ·36930 365 | 18 207 | ·20181 686 | 6 255 |
| ·44 | ·25848 242 | 11 477 | ·59731 200 | 29 107 | ·94 | ·36719 455 | 18 138 | ·21999 069 | 7 197 |
| 4·45 | + 0·26439 767 | − 11 765 | − 0·58569 051 | + 28 708 | 4·95 | + 0·36490 407 | − 18 062 | + 0·23809 255 | − 8 135 |
| ·46 | ·27019 527 | 12 050 | ·57378 197 | 28 295 | ·96 | ·36243 297 | 17 975 | ·25611 305 | 9 079 |
| ·47 | ·27587 237 | 12 332 | ·56159 051 | 27 864 | ·97 | ·35978 212 | 17 882 | ·27404 276 | 10 022 |
| ·48 | ·28142 615 | 12 607 | ·54912 044 | 27 416 | ·98 | ·35695 245 | 17 774 | ·29187 225 | 10 966 |
| ·49 | ·28685 386 | 12 879 | ·53637 624 | 26 954 | ·99 | ·35394 504 | 17 662 | ·30959 208 | 11 909 |
| 4·50 | + 0·29215 278 | − 13 147 | − 0·52336 253 | + 26 474 | 5·00 | + 0·35076 101 | − 17 537 | + 0·32719 282 | − 12 852 |

## TABLE I—Ai(x) AND Ai'(x)

| x | Ai(−x) | $\delta^2$ | Ai'(−x) | $\delta_m^2$ | x | Ai(−x) | $\delta^2$ | Ai'(−x) | $\delta_m^2$ |
|---|---|---|---|---|---|---|---|---|---|
| 5·00 | +0·35076 101 | − 17 537 | +0·32719 282 | − 12 852 | 5·50 | +0·01778 154 | − 977 | +0·86419 722 | − 47 356 |
| ·01 | ·34740 161 | 17 404 | ·34466 504 | 13 795 | ·51 | ·00913 547 | 503 | ·86493 807 | 47 568 |
| ·02 | ·34386 817 | 17 262 | ·36199 932 | 14 733 | ·52 | + ·00048 437 | − 27 | ·86520 328 | 47 758 |
| ·03 | ·34016 211 | 17 109 | ·37918 627 | 15 672 | ·53 | − ·00816 700 | + 451 | ·86499 096 | 47 918 |
| ·04 | ·33628 496 | 16 947 | ·39621 651 | 16 606 | ·54 | ·01681 386 | 933 | ·86429 951 | 48 053 |
| 5·05 | +0·33223 834 | − 16 778 | +0·41308 069 | − 17 540 | 5·55 | −0·02545 139 | + 1 412 | +0·86312 758 | − 48 161 |
| ·06 | ·32802 394 | 16 598 | ·42976 948 | 18 465 | ·56 | ·03407 480 | 1 894 | ·86147 409 | 48 241 |
| ·07 | ·32364 356 | 16 407 | ·44627 362 | 19 390 | ·57 | ·04267 927 | 2 378 | ·85933 824 | 48 295 |
| ·08 | ·31909 911 | 16 210 | ·46258 387 | 20 309 | ·58 | ·05125 996 | 2 860 | ·85671 949 | 48 320 |
| ·09 | ·31439 256 | 16 001 | ·47869 104 | 21 222 | ·59 | ·05981 205 | 3 344 | ·85361 759 | 48 318 |
| 5·10 | +0·30952 600 | − 15 786 | +0·49458 600 | − 22 129 | 5·60 | −0·06833 070 | + 3 826 | +0·85003 256 | − 48 288 |
| ·11 | ·30450 158 | 15 559 | ·51025 968 | 23 030 | ·61 | ·07681 109 | 4 309 | ·84596 470 | 48 229 |
| ·12 | ·29932 157 | 15 324 | ·52570 307 | 23 923 | ·62 | ·08524 839 | 4 791 | ·84141 460 | 48 144 |
| ·13 | ·29398 832 | 15 082 | ·54090 724 | 24 810 | ·63 | ·09363 778 | 5 272 | ·83638 312 | 48 027 |
| ·14 | ·28850 425 | 14 827 | ·55586 333 | 25 686 | ·64 | ·10197 445 | 5 751 | ·83087 142 | 47 885 |
| 5·15 | +0·28287 191 | − 14 568 | +0·57056 257 | − 26 557 | 5·65 | −0·11025 361 | + 6 229 | +0·82488 093 | − 47 711 |
| ·16 | ·27709 389 | 14 298 | ·58499 626 | 27 416 | ·66 | ·11847 048 | 6 706 | ·81841 338 | 47 509 |
| ·17 | ·27117 289 | 14 018 | ·59915 581 | 28 266 | ·67 | ·12662 029 | 7 178 | ·81147 079 | 47 279 |
| ·18 | ·26511 171 | 13 732 | ·61303 272 | 29 105 | ·68 | ·13469 832 | 7 652 | ·80405 546 | 47 022 |
| ·19 | ·25891 321 | 13 437 | ·62661 860 | 29 933 | ·69 | ·14269 983 | 8 118 | ·79616 997 | 46 731 |
| 5·20 | +0·25258 034 | − 13 134 | +0·63990 517 | − 30 752 | 5·70 | −0·15062 016 | + 8 586 | +0·78781 722 | − 46 414 |
| ·21 | ·24611 613 | 12 822 | ·65288 425 | 31 554 | ·71 | ·15845 463 | 9 047 | ·77900 038 | 46 068 |
| ·22 | ·23952 370 | 12 502 | ·66554 781 | 32 349 | ·72 | ·16619 863 | 9 506 | ·76972 291 | 45 693 |
| ·23 | ·23280 625 | 12 175 | ·67788 791 | 33 126 | ·73 | ·17384 757 | 9 962 | ·75998 856 | 45 290 |
| ·24 | ·22596 705 | 11 840 | ·68989 677 | 33 893 | ·74 | ·18139 689 | 10 411 | ·74980 137 | 44 854 |
| 5·25 | +0·21900 945 | − 11 498 | +0·70156 673 | − 34 644 | 5·75 | −0·18884 210 | + 10 858 | +0·73916 569 | − 44 395 |
| ·26 | ·21193 687 | 11 147 | ·71289 028 | 35 380 | ·76 | ·19617 873 | 11 300 | ·72808 612 | 43 902 |
| ·27 | ·20475 282 | 10 790 | ·72386 006 | 36 102 | ·77 | ·20340 236 | 11 736 | ·71656 758 | 43 382 |
| ·28 | ·19746 087 | 10 425 | ·73446 885 | 36 807 | ·78 | ·21050 863 | 12 166 | ·70461 527 | 42 836 |
| 29 | 19006 467 | 10 054 | 74470 960 | 37 496 | 79 | 21749 384 | 12 593 | 69223 466 | 42 258 |
| 5·30 | +0·18256 793 | − 9 675 | +0·75457 542 | − 38 168 | 5·80 | −0·22435 192 | + 13 012 | +0·67943 152 | − 41 653 |
| ·31 | ·17497 444 | 9 292 | ·76405 959 | 38 825 | ·81 | ·23108 048 | 13 425 | ·66621 190 | 41 022 |
| ·32 | ·16728 803 | 8 898 | ·77315 555 | 39 460 | ·82 | ·23767 479 | 13 832 | ·65258 212 | 40 359 |
| ·33 | ·15951 264 | 8 502 | ·78185 694 | 40 079 | ·83 | ·24413 078 | 14 233 | ·63854 880 | 39 671 |
| ·34 | ·15165 223 | 8 098 | ·79015 757 | 40 681 | ·84 | ·25044 444 | 14 625 | ·62411 882 | 38 958 |
| 5·35 | +0·14371 084 | − 7 687 | +0·79805 143 | − 41 261 | 5·85 | −0·25661 185 | + 15 011 | +0·60929 932 | − 38 211 |
| ·36 | ·13569 258 | 7 274 | ·80553 272 | 41 820 | ·86 | ·26262 915 | 15 389 | ·59409 775 | 37 443 |
| ·37 | ·12760 158 | 6 851 | ·81259 584 | 42 363 | ·87 | ·26849 256 | 15 760 | ·57852 180 | 36 648 |
| ·38 | ·11944 207 | 6 425 | ·81923 537 | 42 883 | ·88 | ·27419 837 | 16 123 | ·56257 942 | 35 823 |
| ·39 | ·11121 831 | 5 995 | ·82544 611 | 43 382 | ·89 | ·27974 295 | 16 475 | ·54627 885 | 34 976 |
| 5·40 | +0·10293 460 | − 5 558 | +0·83122 307 | − 43 857 | 5·90 | −0·28512 278 | + 16 822 | +0·52962 857 | − 34 102 |
| ·41 | ·09459 531 | 5 118 | ·83656 149 | 44 316 | ·91 | ·29033 439 | 17 158 | ·51263 732 | 33 202 |
| ·42 | ·08620 484 | 4 671 | ·84145 680 | 44 745 | ·92 | ·29537 442 | 17 485 | ·49531 409 | 32 280 |
| ·43 | ·07776 766 | 4 223 | ·84590 469 | 45 159 | ·93 | ·30023 960 | 17 804 | ·47766 811 | 31 330 |
| ·44 | ·06928 825 | 3 768 | ·84990 104 | 45 543 | ·94 | ·30492 674 | 18 112 | ·45970 887 | 30 358 |
| 5·45 | +0·06077 116 | − 3 313 | +0·85344 200 | − 45 908 | 5·95 | −0·30943 276 | + 18 410 | +0·44144 609 | − 29 362 |
| ·46 | ·05222 094 | 2 850 | ·85652 393 | 46 246 | ·96 | ·31375 468 | 18 699 | ·42288 973 | 28 344 |
| ·47 | ·04364 222 | 2 387 | ·85914 344 | 46 562 | ·97 | ·31788 961 | 18 977 | ·40404 997 | 27 303 |
| ·48 | ·03503 963 | 1 920 | ·86129 738 | 46 850 | ·98 | ·32183 477 | 19 245 | ·38493 722 | 26 240 |
| ·49 | ·02641 784 | 1 451 | ·86298 286 | 47 117 | ·99 | ·32558 748 | 19 502 | ·36556 211 | 25 155 |
| 5·50 | +0·01778 154 | − 977 | +0·86419 722 | − 47 356 | 6·00 | −0·32914 517 | + 19 747 | +0·34593 549 | − 24 050 |

TABLE I—Ai(x) AND Ai'(x)

| $x$ | Ai$(-x)$ | $\delta^2_m$ | Ai$'(-x)$ | $\delta^2_m$ | $x$ | Ai$(-x)$ | $\delta^2_m$ | Ai$'(-x)$ | $\delta^2_m$ |
|---|---|---|---|---|---|---|---|---|---|
| 6·00 | −0·32914 517 | +19 749 | +0·34593 549 | −24 050 | 6·50 | −0·23802 030 | +15 472 | −0·67495 249 | +41 494 |
| ·01 | ·33250 539 | 19 985 | ·32606 841 | 22 923 | ·51 | ·23119 412 | 15 054 | ·69021 446 | 42 623 |
| ·02 | ·33566 578 | 20 208 | ·30597 213 | 21 779 | ·52 | ·22421 743 | 14 618 | ·70505 024 | 43 729 |
| ·03 | ·33862 411 | 20 422 | ·28565 810 | 20 613 | ·53 | ·21709 457 | 14 178 | ·71944 877 | 44 812 |
| ·04 | ·34137 825 | 20 618 | ·26513 797 | 19 429 | ·54 | ·20982 995 | 13 724 | ·73339 923 | 45 869 |
| 6·05 | −0·34392 622 | +20 810 | +0·24442 358 | −18 230 | 6·55 | −0·20242 811 | +13 260 | −0·74689 105 | +46 900 |
| ·06 | ·34626 612 | 20 985 | ·22352 693 | 17 008 | ·56 | ·19489 369 | 12 786 | ·75991 392 | 47 904 |
| ·07 | ·34839 619 | 21 148 | ·20246 022 | 15 776 | ·57 | ·18723 143 | 12 301 | ·77245 780 | 48 881 |
| ·08 | ·35031 480 | 21 302 | ·18123 578 | 14 524 | ·58 | ·17944 617 | 11 810 | ·78451 292 | 49 829 |
| ·09 | ·35202 042 | 21 438 | ·15986 613 | 13 256 | ·59 | ·17154 283 | 11 304 | ·79606 980 | 50 749 |
| 6·10 | −0·35351 168 | +21 567 | +0·13836 394 | −11 978 | 6·60 | −0·16352 646 | +10 794 | −0·80711 925 | +51 638 |
| ·11 | ·35478 730 | 21 678 | ·11674 200 | 10 681 | ·61 | ·15540 217 | 10 274 | ·81765 238 | 52 496 |
| ·12 | ·35584 616 | 21 780 | ·09501 327 | 9 376 | ·62 | ·14717 516 | 9 742 | ·82766 061 | 53 323 |
| ·13 | ·35668 725 | 21 865 | ·07319 081 | 8 054 | ·63 | ·13885 074 | 9 208 | ·83713 567 | 54 117 |
| ·14 | ·35730 971 | 21 941 | ·05128 783 | 6 725 | ·64 | ·13043 426 | 8 663 | ·84606 962 | 54 877 |
| 6·15 | −0·35771 279 | +22 002 | +0·02931 763 | − 5 379 | 6·65 | −0·12193 117 | + 8 108 | −0·85445 486 | +55 607 |
| ·16 | ·35789 588 | 22 047 | + ·00729 365 | 4 032 | ·66 | ·11334 701 | 7 549 | ·86228 410 | 56 297 |
| ·17 | ·35785 852 | 22 080 | − ·01477 062 | 2 667 | ·67 | ·10468 737 | 6 985 | ·86955 043 | 56 955 |
| ·18 | ·35760 038 | 22 102 | ·03686 155 | − 1 298 | ·68 | ·09595 790 | 6 410 | ·87624 727 | 57 580 |
| ·19 | ·35712 125 | 22 108 | ·05896 545 | + 78 | ·69 | ·08716 434 | 5 832 | ·88236 839 | 58 160 |
| 6·20 | −0·35642 107 | +22 098 | −0·08106 856 | + 1 462 | 6·70 | −0·07831 247 | + 5 247 | −0·88790 797 | +58 713 |
| ·21 | ·35549 993 | 22 079 | ·10315 704 | 2 850 | ·71 | ·06940 814 | 4 658 | ·89286 050 | 59 220 |
| ·22 | ·35435 803 | 22 043 | ·12521 701 | 4 244 | ·72 | ·06045 724 | 4 064 | ·89722 090 | 59 692 |
| ·23 | ·35299 573 | 21 992 | ·14723 453 | 5 644 | ·73 | ·05146 571 | 3 464 | ·90098 445 | 60 127 |
| ·24 | ·35141 353 | 21 931 | ·16919 561 | 7 042 | ·74 | ·04243 955 | 2 861 | ·90414 681 | 60 519 |
| 6·25 | −0·34961 205 | +21 851 | −0·19108 626 | + 8 447 | 6·75 | −0·03338 479 | + 2 254 | −0·90670 405 | +60 872 |
| ·26 | ·34759 208 | 21 761 | ·21289 244 | 9 851 | ·76 | ·02430 750 | 1 644 | ·90865 264 | 61 185 |
| ·27 | ·34535 453 | 21 656 | ·23460 011 | 11 258 | ·77 | ·01521 378 | 1 031 | ·90998 945 | 61 459 |
| ·28 | ·34290 045 | 21 534 | ·25619 521 | 12 659 | ·78 | − ·00610 976 | + 414 | ·91071 175 | 61 690 |
| ·29 | ·34023 105 | 21 403 | ·27766 372 | 14 062 | ·79 | + ·00299 840 | − 204 | ·91081 723 | 61 877 |
| 6·30 | −0·33734 765 | +21 255 | −0·29899 161 | +15 464 | 6·80 | +0·01210 452 | − 823 | −0·91030 401 | +62 027 |
| ·31 | ·33425 173 | 21 093 | ·32016 487 | 16 861 | ·81 | ·02120 241 | 1 442 | ·90917 060 | 62 131 |
| ·32 | ·33094 491 | 20 916 | ·34116 953 | 18 253 | ·82 | ·03028 587 | 2 068 | ·90741 596 | 62 193 |
| ·33 | ·32742 895 | 20 727 | ·36199 167 | 19 640 | ·83 | ·03934 866 | 2 687 | ·90503 947 | 62 212 |
| ·34 | ·32370 574 | 20 526 | ·38261 742 | 21 023 | ·84 | ·04838 458 | 3 309 | ·90204 094 | 62 188 |
| 6·35 | −0·31977 730 | +20 306 | −0·40303 296 | +22 394 | 6·85 | +0·05738 741 | − 3 933 | −0·89842 061 | +62 120 |
| ·36 | ·31564 582 | 20 077 | ·42322 457 | 23 762 | ·86 | ·06635 092 | 4 551 | ·89417 916 | 62 008 |
| ·37 | ·31131 360 | 19 831 | ·44317 858 | 25 119 | ·87 | ·07526 892 | 5 170 | ·88931 771 | 61 855 |
| ·38 | ·30678 309 | 19 575 | ·46288 142 | 26 465 | ·88 | ·08413 522 | 5 791 | ·88383 780 | 61 653 |
| ·39 | ·30205 686 | 19 304 | ·48231 963 | 27 801 | ·89 | ·09294 362 | 6 401 | ·87774 144 | 61 410 |
| 6·40 | −0·29713 762 | +19 017 | −0·50147 985 | +29 124 | 6·90 | +0·10168 800 | − 7 018 | −0·87103 106 | +61 123 |
| ·41 | ·29202 823 | 18 721 | ·52034 885 | 30 436 | ·91 | ·11036 221 | 7 627 | ·86370 953 | 60 790 |
| ·42 | ·28673 166 | 18 409 | ·53891 352 | 31 731 | ·92 | ·11896 016 | 8 231 | ·85578 018 | 60 414 |
| ·43 | ·28125 102 | 18 087 | ·55716 090 | 33 015 | ·93 | ·12747 580 | 8 835 | ·84724 677 | 59 993 |
| ·44 | ·27558 954 | 17 748 | ·57507 816 | 34 281 | ·94 | ·13590 310 | 9 433 | ·83811 351 | 59 528 |
| 6·45 | −0·26975 060 | +17 402 | −0·59265 264 | +35 530 | 6·95 | +0·14423 608 | − 10 024 | −0·82838 505 | +59 020 |
| ·46 | ·26373 767 | 17 036 | ·60987 185 | 36 762 | ·96 | ·15246 883 | 10 612 | ·81806 647 | 58 467 |
| ·47 | ·25755 439 | 16 667 | ·62672 347 | 37 976 | ·97 | ·16059 547 | 11 196 | ·80716 330 | 57 869 |
| ·48 | ·25120 447 | 16 279 | ·64319 537 | 39 169 | ·98 | ·16861 017 | 11 769 | ·79568 152 | 57 229 |
| ·49 | ·24469 178 | 15 881 | ·65927 562 | 40 342 | ·99 | ·17650 719 | 12 338 | ·78362 753 | 56 545 |
| 6·50 | −0·23802 030 | +15 472 | −0·67495 249 | +41 494 | 7·00 | +0·18428 084 | − 12 902 | −0·77100 817 | +55 818 |

TABLE I—Ai(x) AND Ai'(x)

| $x$ | Ai($-x$) | $\delta^2_m$ | Ai'($-x$) | $\delta^2_m$ | $x$ | Ai($-x$) | $\delta^2_m$ | Ai'($-x$) | $\delta^2_m$ |
|---|---|---|---|---|---|---|---|---|---|
| 7·00 | +0·18428 084 | − 12 902 | −0·77100 817 | +55 818 | 7·50 | +0·32177 572 | −24 136 | +0·31880 951 | −20 695 |
| ·01 | ·19192 549 | 13 454 | ·75783 071 | 55 048 | ·51 | ·31846 731 | 23 918 | ·34283 610 | 22 562 |
| ·02 | ·19943 561 | 14 000 | ·74410 285 | 54 233 | ·52 | ·31491 975 | 23 685 | ·36663 708 | 24 423 |
| ·03 | ·20680 574 | 14 540 | ·72983 273 | 53 380 | ·53 | ·31113 538 | 23 429 | ·39019 385 | 26 271 |
| ·04 | ·21403 049 | 15 069 | ·71502 889 | 52 484 | ·54 | ·30711 675 | 23 160 | ·41348 793 | 28 108 |
| 7·05 | +0·22110 457 | − 15 589 | −0·69970 029 | +51 542 | 7·55 | +0·30286 656 | −22 868 | +0·43650 096 | −29 929 |
| ·06 | ·22802 278 | 16 100 | ·68385 634 | 50 566 | ·56 | ·29838 772 | 22 559 | ·45921 473 | 31 734 |
| ·07 | ·23478 001 | 16 600 | ·66750 681 | 49 544 | ·57 | ·29368 332 | 22 233 | ·48161 119 | 33 522 |
| ·08 | ·24137 126 | 17 090 | ·65066 191 | 48 484 | ·58 | ·28875 662 | 21 891 | ·50367 246 | 35 294 |
| ·09 | ·24779 163 | 17 569 | ·63333 224 | 47 385 | ·59 | ·28361 105 | 21 528 | ·52538 083 | 37 042 |
| 7·10 | +0·25403 633 | − 18 038 | −0·61552 879 | +46 247 | 7·60 | +0·27825 023 | −21 148 | +0·54671 882 | −38 770 |
| ·11 | ·26010 067 | 18 494 | ·59726 294 | 45 070 | ·61 | ·27267 796 | 20 752 | ·56766 915 | 40 477 |
| ·12 | ·26598 009 | 18 939 | ·57854 646 | 43 857 | ·62 | ·26689 820 | 20 340 | ·58821 476 | 42 154 |
| ·13 | ·27167 014 | 19 373 | ·55939 148 | 42 603 | ·63 | ·26091 507 | 19 909 | ·60833 887 | 43 809 |
| ·14 | ·27716 649 | 19 788 | ·53981 053 | 41 319 | ·64 | ·25473 288 | 19 463 | ·62802 494 | 45 439 |
| 7·15 | +0·28246 497 | − 20 199 | −0·51981 646 | +39 995 | 7·65 | +0·24835 609 | − 19 001 | +0·64725 668 | −47 033 |
| ·16 | ·28756 149 | 20 590 | ·49942 250 | 38 637 | ·66 | ·24178 932 | 18 523 | ·66601 814 | 48 603 |
| ·17 | ·29245 213 | 20 971 | ·47864 223 | 37 246 | ·67 | ·23503 735 | 18 029 | ·68429 363 | 50 136 |
| ·18 | ·29713 309 | 21 334 | ·45748 956 | 35 823 | ·68 | ·22810 512 | 17 521 | ·70206 781 | 51 643 |
| ·19 | ·30160 073 | 21 688 | ·43597 872 | 34 366 | ·69 | ·22099 771 | 16 995 | ·71932 563 | 53 109 |
| 7·20 | +0·30585 152 | − 22 021 | −0·41412 428 | +32 879 | 7·70 | +0·21372 037 | − 16 458 | +0·73605 242 | −54 543 |
| ·21 | ·30988 212 | 22 345 | ·39194 111 | 31 360 | ·71 | ·20627 848 | 15 906 | ·75223 385 | 55 938 |
| ·22 | ·31368 930 | 22 650 | ·36944 439 | 29 815 | ·72 | ·19867 756 | 15 340 | ·76785 597 | 57 295 |
| ·23 | ·31727 001 | 22 940 | ·34664 958 | 28 238 | ·73 | ·19092 327 | 14 759 | ·78290 521 | 58 615 |
| ·24 | ·32062 135 | 23 215 | ·32357 244 | 26 634 | ·74 | ·18302 141 | 14 167 | ·79736 838 | 59 888 |
| 7·25 | +0·32374 057 | − 23 472 | −0·30022 900 | +25 008 | 7·75 | +0·17497 790 | − 13 562 | +0·81123 274 | −61 126 |
| ·26 | ·32662 510 | 23 715 | ·27663 553 | 23 352 | ·76 | ·16679 879 | 12 944 | ·82448 592 | 62 317 |
| ·27 | ·32927 251 | 23 940 | ·25280 858 | 21 673 | ·77 | ·15849 026 | 12 316 | ·83711 601 | 63 463 |
| ·28 | ·33168 055 | 24 148 | ·22876 494 | 19 975 | ·78 | ·15005 859 | 11 676 | ·84911 155 | 64 564 |
| ·29 | ·33384 711 | 24 339 | ·30153 160 | 18 350 | ·79 | ·14151 018 | 11 025 | ·86046 153 | 65 620 |
| 7·30 | +0·33577 037 | − 24 513 | −0·18009 580 | +16 507 | 7·80 | +0·13285 154 | − 10 363 | +0·87115 540 | −66 626 |
| ·31 | ·33744 850 | 24 668 | ·15550 497 | 14 743 | ·81 | ·12408 929 | 9 691 | ·88118 310 | 67 586 |
| ·32 | ·33887 998 | 24 808 | ·13076 674 | 12 963 | ·82 | ·11523 014 | 9 015 | ·89053 504 | 68 492 |
| ·33 | ·34006 341 | 24 930 | ·10589 891 | 11 165 | ·83 | ·10628 087 | 8 321 | ·89920 215 | 69 349 |
| ·34 | ·34099 758 | 25 028 | ·08091 946 | 9 351 | ·84 | ·09724 840 | 7 624 | ·90717 586 | 70 157 |
| 7·35 | +0·34168 149 | − 25 119 | −0·05584 653 | + 7 522 | 7·85 | +0·08813 970 | − 6 922 | +0·91444 810 | −70 906 |
| ·36 | ·34211 426 | 25 180 | ·03069 840 | 5 682 | ·86 | ·07896 180 | 6 204 | ·92101 137 | 71 608 |
| ·37 | ·34229 526 | 25 228 | − ·00549 347 | 3 829 | ·87 | ·06972 186 | 5 490 | ·92685 866 | 72 252 |
| ·38 | ·34222 401 | 25 260 | + ·01974 973 | 1 966 | ·88 | ·06042 704 | 4 762 | ·93198 353 | 72 841 |
| ·39 | ·34190 020 | 25 267 | ·04501 257 | + 92 | ·89 | ·05108 461 | 4 031 | ·93638 009 | 73 375 |
| 7·40 | +0·34132 375 | − 25 259 | +0·07027 632 | − 1 786 | 7·90 | +0·04170 188 | − 3 295 | +0·94004 300 | −73 853 |
| ·41 | ·34049 474 | 25 234 | ·09552 220 | 3 672 | ·91 | ·03228 621 | 2 553 | ·94296 749 | 74 270 |
| ·42 | ·33941 343 | 25 185 | ·12073 135 | 5 565 | ·92 | ·02284 501 | 1 811 | ·94514 938 | 74 636 |
| ·43 | ·33808 030 | 25 123 | ·14588 485 | 7 459 | ·93 | ·01338 571 | 1 060 | ·94658 503 | 74 934 |
| ·44 | ·33649 598 | 25 036 | ·17096 376 | 9 353 | ·94 | + ·00391 581 | − 313 | ·94727 144 | 75 182 |
| 7·45 | +0·33466 133 | − 24 935 | +0·19594 913 | −11 253 | 7·95 | −0·00555 721 | + 442 | +0·94720 615 | −75 364 |
| ·46 | ·33257 737 | 24 811 | ·22082 198 | 13 147 | ·96 | ·01502 581 | 1 197 | ·94638 733 | 75 489 |
| ·47 | ·33024 533 | 24 672 | ·24556 336 | 15 043 | ·97 | ·02448 244 | 1 951 | ·94481 373 | 75 554 |
| ·48 | ·32766 661 | 24 511 | ·27015 432 | 16 929 | ·98 | ·03391 956 | 2 705 | ·94248 471 | 75 553 |
| ·49 | ·32484 281 | 24 332 | ·29457 599 | 18 816 | ·99 | ·04332 962 | 3 464 | ·93940 026 | 75 499 |
| 7·50 | +0·32177 572 | − 24 136 | +0·31880 951 | −20 695 | 8·00 | −0·05270 505 | + 4 217 | +0·93556 094 | −75 379 |

## TABLE I—Ai($x$) AND Ai'($x$)

| $x$ | Ai($-x$) | $\delta^2_m$ | Ai'($-x$) | $\delta^2_m$ | $x$ | Ai($-x$) | $\delta^2_m$ | Ai'($-x$) | $\delta^2_m$ |
|---|---|---|---|---|---|---|---|---|---|
| 8·00 | −0·05270 505 | + 4 217 | +0·93556 094 | − 75 379 | 8·50 | −0·33029 024 | +28 077 | −0·03231 335 | −    556 |
| ·01 | ·06203 832 | 4 970 | ·93096 795 | 75 196 | ·51 | ·32982 673 | 28 070 | ·06038 681 | +  1 840 |
| ·02 | ·07132 190 | 5 719 | ·92562 311 | 74 956 | ·52 | ·32908 256 | 28 039 | ·08844 186 | 4 244 |
| ·03 | ·08054 829 | 6 470 | ·91952 883 | 74 649 | ·53 | ·32805 804 | 27 987 | ·11645 446 | 6 654 |
| ·04 | ·08970 999 | 7 213 | ·91268 817 | 74 283 | ·54 | ·32675 370 | 27 907 | ·14440 052 | 9 062 |
| 8·05 | −0·09879 957 | + 7 952 | +0·90510 479 | − 73 855 | 8·55 | −0·32517 033 | +27 804 | −0·17225 595 | +11 478 |
| ·06 | ·10780 963 | 8 691 | ·89678 297 | 73 366 | ·56 | ·32330 896 | 27 677 | ·19999 661 | 13 886 |
| ·07 | ·11673 279 | 9 421 | ·88772 761 | 72 812 | ·57 | ·32117 086 | 27 528 | ·22759 841 | 16 294 |
| ·08 | ·12556 175 | 10 145 | ·87794 424 | 72 201 | ·58 | ·31875 753 | 27 349 | ·25503 728 | 18 695 |
| ·09 | ·13428 927 | 10 866 | ·86743 898 | 71 524 | ·59 | ·31607 074 | 27 155 | ·28228 921 | 21 089 |
| 8·10 | −0·14290 815 | +11 576 | +0·85621 859 | − 70 789 | 8·60 | −0·31311 245 | +26 929 | −0·30933 027 | +23 472 |
| ·11 | ·15141 128 | 12 281 | ·84429 042 | 69 994 | ·61 | ·30988 491 | 26 683 | ·33613 663 | 25 845 |
| ·12 | ·15979 162 | 12 975 | ·83166 243 | 69 134 | ·62 | ·30639 058 | 26 413 | ·36268 457 | 28 199 |
| ·13 | ·16804 222 | 13 663 | ·81834 321 | 68 218 | ·63 | ·30263 216 | 26 119 | ·38895 054 | 30 544 |
| ·14 | ·17615 621 | 14 340 | ·80434 192 | 67 239 | ·64 | ·29861 259 | 25 804 | ·41491 111 | 32 863 |
| 8·15 | −0·18412 682 | +15 008 | +0·78966 834 | − 66 208 | 8·65 | −0·29433 503 | +25 460 | −0·44054 308 | +35 165 |
| ·16 | ·19194 737 | 15 662 | ·77433 280 | 65 109 | ·66 | ·28980 290 | 25 101 | ·46582 344 | 37 446 |
| ·17 | ·19961 131 | 16 311 | ·75834 627 | 63 960 | ·67 | ·28501 981 | 24 713 | ·49072 939 | 39 697 |
| ·18 | ·20711 217 | 16 943 | ·74172 025 | 62 749 | ·68 | ·27998 963 | 24 305 | ·51523 841 | 41 927 |
| ·19 | ·21444 362 | 17 565 | ·72446 684 | 61 483 | ·69 | ·27471 644 | 23 875 | ·53932 822 | 44 124 |
| 8·20 | −0·22159 945 | +18 171 | +0·70659 870 | − 60 163 | 8·70 | −0·26920 454 | +23 423 | −0·56297 685 | +46 290 |
| ·21 | ·22857 359 | 18 768 | ·68812 903 | 58 787 | ·71 | ·26345 845 | 22 950 | ·58616 264 | 48 423 |
| ·22 | ·23536 008 | 19 349 | ·66907 159 | 57 355 | ·72 | ·25748 290 | 22 453 | ·60886 426 | 50 522 |
| ·23 | ·24195 311 | 19 913 | ·64944 069 | 55 875 | ·73 | ·25128 285 | 21 940 | ·63106 073 | 52 584 |
| ·24 | ·24834 703 | 20 466 | ·62925 114 | 54 338 | ·74 | ·24486 344 | 21 402 | ·65273 144 | 54 603 |
| 8·25 | −0·25453 632 | +20 999 | +0·60851 830 | − 52 755 | 8·75 | −0·23823 004 | +20 848 | −0·67385 619 | +56 585 |
| ·26 | ·26051 564 | 21 521 | ·58725 801 | 51 115 | ·76 | ·23138 820 | 20 270 | ·69441 517 | 58 521 |
| ·27 | ·26627 978 | 22 023 | ·56548 665 | 49 436 | ·77 | ·22434 369 | 19 678 | ·71438 902 | 60 414 |
| ·28 | ·27182 372 | 22 508 | ·54322 103 | 47 700 | ·78 | ·21710 244 | 19 065 | ·73375 882 | 62 256 |
| ·29 | ·27714 261 | 22 977 | ·52047 849 | 45 925 | ·79 | ·20967 058 | 18 429 | ·75250 614 | 64 055 |
| 8·30 | −0·28223 176 | +23 428 | +0·49727 679 | − 44 100 | 8·80 | −0·20205 445 | +17 784 | −0·77061 301 | +65 799 |
| ·31 | ·28708 667 | 23 857 | ·47363 417 | 42 235 | ·81 | ·19426 052 | 17 116 | ·78806 199 | 67 491 |
| ·32 | ·29170 304 | 24 273 | ·44956 928 | 40 326 | ·82 | ·18629 546 | 16 433 | ·80483 616 | 69 129 |
| ·33 | ·29607 672 | 24 666 | ·42510 121 | 38 375 | ·83 | ·17816 610 | 15 734 | ·82091 914 | 70 710 |
| ·34 | ·30020 378 | 25 038 | ·40024 946 | 36 387 | ·84 | ·16987 943 | 15 019 | ·83629 512 | 72 238 |
| 8·35 | −0·30408 049 | +25 393 | +0·37503 391 | − 34 358 | 8·85 | −0·16144 260 | +14 288 | −0·85094 884 | +73 698 |
| ·36 | ·30770 331 | 25 726 | ·34947 484 | 32 300 | ·86 | ·15286 291 | 13 546 | ·86486 568 | 75 106 |
| ·37 | ·31106 891 | 26 039 | ·32359 285 | 30 196 | ·87 | ·14414 779 | 12 786 | ·87803 158 | 76 447 |
| ·38 | ·31417 416 | 26 328 | ·29740 895 | 28 070 | ·88 | ·13530 483 | 12 018 | ·89043 313 | 77 724 |
| ·39 | ·31701 616 | 26 602 | ·27094 442 | 25 906 | ·89 | ·12634 172 | 11 231 | ·90205 756 | 78 935 |
| 8·40 | −0·31959 219 | +26 848 | +0·24422 089 | − 23 711 | 8·90 | −0·11726 631 | +10 440 | −0·91289 276 | +80 083 |
| ·41 | ·32189 978 | 27 073 | ·21726 029 | 21 496 | ·91 | ·10808 653 | 9 629 | ·92292 726 | 81 159 |
| ·42 | ·32393 668 | 27 278 | ·19008 479 | 19 244 | ·92 | ·09881 047 | 8 816 | ·93215 030 | 82 166 |
| ·43 | ·32570 084 | 27 459 | ·16271 688 | 16 978 | ·93 | ·08944 627 | 7 989 | ·94055 181 | 83 106 |
| ·44 | ·32719 045 | 27 617 | ·13517 924 | 14 682 | ·94 | ·08000 220 | 7 152 | ·94812 240 | 83 968 |
| 8·45 | −0·32840 393 | +27 752 | +0·10749 481 | − 12 370 | 8·95 | −0·07048 662 | + 6 309 | −0·95485 344 | +84 763 |
| ·46 | ·32933 993 | 27 864 | ·07968 672 | 10 037 | ·96 | ·06090 796 | 5 458 | ·96073 699 | 85 481 |
| ·47 | ·32999 733 | 27 954 | ·05177 829 | 7 685 | ·97 | ·05127 473 | 4 600 | ·96576 587 | 86 124 |
| ·48 | ·33037 524 | 28 017 | + ·02379 303 | 5 324 | ·98 | ·04159 551 | 3 736 | ·96993 365 | 86 691 |
| ·49 | ·33047 302 | 28 061 | − ·00424 544 | 2 946 | ·99 | ·03187 894 | 2 866 | ·97323 466 | 87 184 |
| 8·50 | −0·33029 024 | +28 077 | −0·03231 335 | −    556 | 9·00 | −0·02213 372 | + 1 993 | −0·97566 398 | +87 596 |

## Table I—Ai($x$) and Ai$'(x)$

| $x$ | Ai($-x$) | $\delta^2_m$ | Ai$'(-x)$ | $\delta^2_m$ | $x$ | Ai($-x$) | $\delta^2_m$ | Ai$'(-x)$ | $\delta^2_m$ |
|---|---|---|---|---|---|---|---|---|---|
| 9·00 | −0·02213 372 | + 1 993 | −0·97566 398 | +87 596 | 9·50 | +0·31910 325 | − 30 319 | −0·10809 532 | +13 464 |
| ·01 | ·01236 858 | 1 115 | ·97721 748 | 87 932 | ·51 | ·32003 241 | 30 437 | ·07771 798 | 10 592 |
| ·02 | − ·00259 230 | + 233 | ·97789 181 | 88 188 | ·52 | ·32065 725 | 30 529 | ·04723 475 | 7 705 |
| ·03 | + ·00718 631 | − 647 | ·97768 441 | 88 365 | ·53 | ·32097 685 | 30 593 | − ·01667 450 | 4 801 |
| ·04 | ·01695 844 | 1 533 | ·97659 351 | 88 462 | ·54 | ·32099 058 | 30 624 | + ·01393 373 | + 1 882 |
| 9·05 | +0·02671 524 | 2 420 | −0·97461 814 | +88 479 | 9·55 | +0·32069 812 | − 30 632 | +0·04456 076 | − 1 049 |
| ·06 | ·03644 785 | 3 301 | ·97175 813 | 88 414 | ·56 | ·32009 940 | 30 603 | ·07517 729 | 3 985 |
| ·07 | ·04614 745 | 4 185 | ·96801 413 | 88 268 | ·57 | ·31919 470 | 30 549 | ·10575 396 | 6 927 |
| ·08 | ·05580 520 | 5 069 | ·96338 760 | 88 042 | ·58 | ·31798 456 | 30 468 | ·13626 135 | 9 876 |
| ·09 | ·06541 227 | 5 945 | ·95788 080 | 87 733 | ·59 | ·31646 980 | 30 351 | ·16666 999 | 12 820 |
| 9·10 | +0·07495 989 | − 6 823 | −0·95149 682 | +87 345 | 9·60 | +0·31465 158 | − 30 210 | +0·19695 044 | − 15 761 |
| ·11 | ·08443 929 | 7 691 | ·94423 954 | 86 873 | ·61 | ·31253 132 | 30 038 | ·22707 329 | 18 696 |
| ·12 | ·09384 178 | 8 559 | ·93611 368 | 86 320 | ·62 | ·31011 074 | 29 835 | ·25700 919 | 21 626 |
| ·13 | ·10315 869 | 9 419 | ·92712 477 | 85 686 | ·63 | ·30739 186 | 29 606 | ·28672 886 | 24 538 |
| ·14 | ·11238 142 | 10 272 | ·91727 915 | 84 971 | ·64 | ·30437 698 | 29 344 | ·31620 317 | 27 440 |
| 9·15 | +0·12150 144 | − 11 117 | −0·90658 397 | +84 176 | 9·65 | +0·30106 871 | − 29 057 | +0·34540 311 | − 30 324 |
| ·16 | ·13051 030 | 11 957 | ·89504 718 | 83 300 | ·66 | ·29746 993 | 28 738 | ·37429 985 | 33 182 |
| ·17 | ·13939 961 | 12 784 | ·88267 754 | 82 344 | ·67 | ·29358 382 | 28 394 | ·40286 480 | 36 026 |
| ·18 | ·14816 110 | 13 602 | ·86948 461 | 81 309 | ·68 | ·28941 383 | 28 017 | ·43106 955 | 38 836 |
| ·19 | ·15678 659 | 14 410 | ·85547 874 | 80 195 | ·69 | ·28496 372 | 27 616 | ·45888 599 | 41 620 |
| 9·20 | +0·16526 800 | − 15 205 | −0·84067 107 | +79 001 | 9·70 | +0·28023 750 | − 27 186 | +0·48628 629 | − 44 370 |
| ·21 | ·17359 738 | 15 991 | ·82507 353 | 77 735 | ·71 | ·27523 947 | 26 728 | ·51324 295 | 47 087 |
| ·22 | ·18176 688 | 16 759 | ·80869 879 | 76 385 | ·72 | ·26997 421 | 26 244 | ·53972 881 | 49 767 |
| ·23 | ·18976 881 | 17 518 | ·79156 033 | 74 969 | ·73 | ·26444 656 | 25 734 | ·56571 708 | 52 405 |
| ·24 | ·19759 559 | 18 258 | ·77367 233 | 73 469 | ·74 | ·25866 162 | 25 197 | ·59118 138 | 54 997 |
| 9·25 | +0·20523 981 | − 18 987 | −0·75504 977 | +71 903 | 9·75 | +0·25262 476 | − 24 632 | +0·61609 579 | − 57 550 |
| ·26 | ·21269 419 | 19 697 | ·73570 832 | 70 261 | ·76 | ·24634 162 | 24 046 | ·64043 480 | 60 048 |
| ·27 | ·21995 163 | 20 391 | ·71566 439 | 68 547 | ·77 | ·23981 807 | 23 434 | ·66417 342 | 62 497 |
| ·28 | ·22700 519 | 21 067 | ·69493 511 | 66 768 | ·78 | ·23306 023 | 22 793 | ·68728 717 | 64 891 |
| ·29 | ·23384 811 | 21 726 | ·67353 828 | 64 916 | ·79 | ·22607 449 | 22 137 | ·70973 211 | 67 230 |
| 9·30 | +0·24047 380 | − 22 367 | −0·65149 241 | +63 002 | 9·80 | +0·21886 743 | − 21 449 | +0·73154 486 | − 69 508 |
| ·31 | ·24687 586 | 22 985 | ·62881 665 | 61 016 | ·81 | ·21144 591 | 20 747 | ·75264 264 | 71 727 |
| ·32 | ·25304 810 | 23 587 | ·60553 084 | 58 974 | ·82 | ·20381 697 | 20 017 | ·77302 327 | 73 881 |
| ·33 | ·25898 451 | 24 165 | ·58165 541 | 56 864 | ·83 | ·19598 790 | 19 266 | ·79266 522 | 75 964 |
| ·34 | ·26467 931 | 24 723 | ·55721 145 | 54 696 | ·84 | ·18796 620 | 18 499 | ·81154 765 | 77 986 |
| 9·35 | +0·27012 692 | − 25 261 | −0·53222 064 | +52 469 | 9·85 | +0·17975 955 | − 17 709 | +0·82965 036 | − 79 929 |
| ·36 | ·27532 197 | 25 770 | ·50670 524 | 50 187 | ·86 | ·17137 585 | 16 898 | ·84695 391 | 81 804 |
| ·37 | ·28025 935 | 26 264 | ·48068 807 | 47 847 | ·87 | ·16282 320 | 16 074 | ·86343 956 | 83 600 |
| ·38 | ·28493 414 | 26 730 | ·45419 252 | 45 459 | ·88 | ·15410 985 | 15 228 | ·87908 935 | 85 322 |
| ·39 | ·28934 168 | 27 170 | ·42724 248 | 43 015 | ·89 | ·14524 425 | 14 364 | ·89388 607 | 86 961 |
| 9·40 | +0·29347 756 | − 27 591 | −0·39986 237 | +40 528 | 9·90 | +0·13623 503 | − 13 490 | +0·90781 333 | − 88 519 |
| ·41 | ·29733 758 | 27 981 | ·37207 707 | 37 989 | ·91 | ·12709 094 | 12 597 | ·92085 555 | 89 995 |
| ·42 | ·30091 783 | 28 349 | ·34391 196 | 35 410 | ·92 | ·11782 091 | 11 688 | ·93299 798 | 91 384 |
| ·43 | ·30421 464 | 28 690 | ·31539 283 | 32 789 | ·93 | ·10843 402 | 10 768 | ·94422 673 | 92 685 |
| ·44 | ·30722 460 | 29 007 | ·28654 589 | 30 126 | ·94 | ·09893 947 | 9 836 | ·95452 879 | 93 901 |
| 9·45 | +0·30994 455 | − 29 290 | −0·25739 776 | +27 428 | 9·95 | +0·08934 658 | − 8 891 | +0·96389 201 | − 95 024 |
| ·46 | ·31237 164 | 29 555 | ·22797 542 | 24 693 | ·96 | ·07966 480 | 7 936 | ·97230 516 | 96 053 |
| ·47 | ·31450 324 | 29 786 | ·19830 621 | 21 927 | ·97 | ·06990 368 | 6 971 | ·97975 795 | 96 992 |
| ·48 | ·31633 703 | 29 991 | ·16841 778 | 19 134 | ·98 | ·06007 287 | 5 995 | ·98624 099 | 97 837 |
| ·49 | ·31787 096 | 30 169 | ·13833 807 | 16 308 | ·99 | ·05018 212 | 5 014 | ·99174 584 | 98 583 |
| 9·50 | +0·31910 325 | − 30 319 | −0·10809 532 | +13 464 | 10·00 | +0·04024 124 | − 4 025 | +0·99626 504 | − 99 232 |

## TABLE I—Ai(x) AND Ai'(x)

| x | Ai(−x) | $\delta^2_m$ | Ai'(−x) | $\delta^2_m$ | x | Ai(−x) | $\delta^2_m$ | Ai'(−x) | $\delta^2_m$ |
|---|---|---|---|---|---|---|---|---|---|
| 10·00 | +0·04024 124 | − 4 025 | +0·99626 504 | − 99 232 | 10·50 | −0·31192 604 | +32 758 | +0·09095 749 | − 12 672 |
| ·01 | ·03026 012 | 3 029 | 0·99979 209 | 99 789 | ·51 | ·31267 165 | 32 864 | ·05814 762 | 9 241 |
| ·02 | ·02024 872 | 2 028 | 1·00232 144 | 100 239 | ·52 | ·31308 868 | 32 940 | + ·02524 538 | 5 788 |
| ·03 | ·01021 704 | 1 027 | 1·00384 858 | 100 593 | ·53 | ·31317 637 | 32 982 | − ·00771 471 | − 2 320 |
| ·04 | + ·00017 510 | − 15 | 1·00436 997 | 100 848 | ·54 | ·31293 431 | 32 986 | ·04069 798 | + 1 158 |
| 10·05 | −0·00986 700 | + 990 | +1·00388 307 | −101 000 | 10·55 | −0·31236 245 | +32 959 | −0·07366 965 | + 4 650 |
| ·06 | ·01989 919 | 2 002 | 1·00238 636 | 101 049 | ·56 | ·31146 107 | 32 893 | ·10659 482 | 8 140 |
| ·07 | ·02991 136 | 3 012 | 0·99987 935 | 100 997 | ·57 | ·31023 082 | 32 794 | ·13943 858 | 11 638 |
| ·08 | ·03989 341 | 4 021 | ·99636 256 | 100 843 | ·58 | ·30867 269 | 32 661 | ·17216 597 | 15 128 |
| ·09 | ·04983 525 | 5 029 | ·99183 753 | 100 585 | ·59 | ·30678 801 | 32 494 | ·20474 209 | 18 616 |
| 10·10 | −0·05972 681 | + 6 033 | +0·98630 684 | −100 225 | 10·60 | −0·30457 846 | +32 287 | −0·23713 207 | + 22 091 |
| ·11 | ·06955 805 | 7 033 | ·97977 409 | 99 761 | ·61 | ·30204 610 | 32 052 | ·26930 116 | 25 554 |
| ·12 | ·07931 897 | 8 028 | ·97224 392 | 99 195 | ·62 | ·29919 329 | 31 777 | ·30121 474 | 29 000 |
| ·13 | ·08899 962 | 9 016 | ·96372 199 | 98 526 | ·63 | ·29602 277 | 31 470 | ·33283 836 | 32 421 |
| ·14 | ·09859 012 | 9 997 | ·95421 499 | 97 752 | ·64 | ·29253 761 | 31 131 | ·36413 781 | 35 823 |
| 10·15 | −0·10808 066 | + 10 970 | +0·94373 065 | − 96 880 | 10·65 | −0·28874 121 | +30 754 | −0·39507 909 | + 39 192 |
| ·16 | ·11746 151 | 11 936 | ·93227 770 | 95 903 | ·66 | ·28463 733 | 30 345 | ·42562 851 | 42 530 |
| ·17 | ·12672 302 | 12 887 | ·91986 590 | 94 829 | ·67 | ·28023 006 | 29 906 | ·45575 270 | 45 830 |
| ·18 | ·13585 567 | 13 833 | ·90650 600 | 93 651 | ·68 | ·27552 380 | 29 427 | ·48541 866 | 49 091 |
| ·19 | ·14485 002 | 14 761 | ·89220 977 | 92 374 | ·69 | ·27052 332 | 28 923 | ·51459 379 | 52 311 |
| 10·20 | −0·15369 678 | + 15 677 | +0·87698 998 | − 91 000 | 10·70 | −0·26523 367 | +28 383 | −0·54324 590 | + 55 479 |
| ·21 | ·16238 679 | 16 582 | ·86086 037 | 89 526 | ·71 | ·25966 025 | 27 813 | ·57134 331 | 58 600 |
| ·22 | ·17091 101 | 17 469 | ·84383 567 | 87 960 | ·72 | ·25380 876 | 27 212 | ·59885 482 | 61 666 |
| ·23 | ·17926 057 | 18 340 | ·82593 155 | 86 294 | ·73 | ·24768 521 | 26 579 | ·62574 978 | 64 671 |
| ·24 | ·18742 676 | 19 194 | ·80716 466 | 84 538 | ·74 | ·24129 592 | 25 919 | ·65199 814 | 67 618 |
| 10·25 | −0·19540 104 | + 20 029 | +0·78755 256 | − 82 686 | 10·75 | −0·23464 750 | +25 227 | −0·67757 044 | + 70 501 |
| ·26 | ·20317 506 | 20 849 | ·76711 376 | 80 746 | ·76 | ·22774 686 | 24 508 | ·70243 786 | 73 311 |
| ·27 | ·21074 063 | 21 644 | ·74586 766 | 78 717 | ·77 | ·22060 119 | 23 762 | ·72657 230 | 76 052 |
| ·28 | ·21808 979 | 22 422 | ·72383 455 | 76 600 | ·78 | ·21321 795 | 22 988 | ·74994 635 | 78 722 |
| ·29 | ·22521 477 | 23 177 | ·70103 560 | 74 397 | ·79 | ·20560 488 | 22 186 | ·77253 333 | 81 306 |
| 10·30 | −0·23210 802 | + 23 909 | +0·67749 283 | − 72 111 | 10·80 | −0·19776 999 | +21 362 | −0·79430 739 | + 83 819 |
| ·31 | ·23876 222 | 24 618 | ·65322 910 | 69 744 | ·81 | ·18972 153 | 20 511 | ·81524 343 | 86 237 |
| ·32 | ·24517 028 | 25 305 | ·62826 808 | 67 296 | ·82 | ·18146 800 | 19 637 | ·83531 725 | 88 577 |
| ·33 | ·25132 534 | 25 965 | ·60263 424 | 64 773 | ·83 | ·17301 814 | 18 740 | ·85450 547 | 90 822 |
| ·34 | ·25722 080 | 26 599 | ·57635 281 | 62 176 | ·84 | ·16438 092 | 17 822 | ·87278 564 | 92 975 |
| 10·35 | −0·26285 032 | + 27 208 | +0·54944 976 | − 59 502 | 10·85 | −0·15556 552 | +16 879 | −0·89013 623 | + 95 034 |
| ·36 | ·26820 781 | 27 787 | ·52195 181 | 56 764 | ·86 | ·14658 136 | 15 922 | ·90653 666 | 96 993 |
| ·37 | ·27328 747 | 28 345 | ·49388 635 | 53 956 | ·87 | ·13743 802 | 14 940 | ·92196 734 | 98 856 |
| ·38 | ·27808 374 | 28 867 | ·46528 145 | 51 082 | ·88 | ·12814 531 | 13 946 | ·93640 966 | 100 609 |
| ·39 | ·28259 139 | 29 363 | ·43616 584 | 48 150 | ·89 | ·11871 318 | 12 929 | ·94984 608 | 102 263 |
| 10·40 | −0·28680 546 | + 29 832 | +0·40656 884 | − 45 157 | 10·90 | −0·10915 179 | +11 899 | −0·96226 007 | +103 806 |
| ·41 | ·29072 127 | 30 268 | ·37652 037 | 42 107 | ·91 | ·09947 144 | 10 853 | ·97363 620 | 105 239 |
| ·42 | ·29433 446 | 30 673 | ·34605 092 | 39 008 | ·92 | ·08968 258 | 9 796 | ·98396 014 | 106 563 |
| ·43 | ·29764 098 | 31 047 | ·31519 149 | 35 854 | ·93 | ·07979 579 | 8 723 | 0·99321 866 | 107 774 |
| ·44 | ·30063 709 | 31 389 | ·28397 360 | 32 660 | ·94 | ·06982 179 | 7 639 | 1·00139 966 | 108 866 |
| 10·45 | −0·30331 937 | + 31 701 | +0·25242 920 | − 29 415 | 10·95 | −0·05977 142 | + 6 546 | −1·00849 221 | +109 844 |
| ·46 | ·30568 470 | 31 978 | ·22059 072 | 26 135 | ·96 | ·04965 561 | 5 444 | ·01448 654 | 110 703 |
| ·47 | ·30773 031 | 32 222 | ·18849 096 | 22 815 | ·97 | ·03948 538 | 4 332 | ·01937 406 | 111 444 |
| ·48 | ·30945 376 | 32 436 | ·15616 311 | 19 464 | ·98 | ·02927 184 | 3 214 | ·02314 737 | 112 060 |
| ·49 | ·31085 292 | 32 609 | ·12364 068 | 16 081 | ·99 | ·01902 617 | 2 092 | ·02580 030 | 112 559 |
| 10·50 | −0·31192 604 | + 32 758 | +0·09095 749 | − 12 672 | 11·00 | −0·00875 959 | + 963 | −1·02732 787 | +112 929 |

TABLE I—Ai(x) AND Ai'(x)

| x | Ai(−x) | $\delta_m^2$ | Ai'(−x) | $\delta_m^2$ | x | Ai(−x) | $\delta_m^2$ | Ai'(−x) | $\delta_m^2$ |
|---|---|---|---|---|---|---|---|---|---|
| 11·00 | −0·00875 959 | + 963 | − 1·02732 787 | +112 929 | 11·50 | +0·30542 297 | −35 129 | +0·08772 415 | − 7 033 |
| ·01 | + ·00151 662 | − 165 | ·02772 637 | 113 181 | ·51 | ·30437 024 | 35 036 | ·12280 586 | 11 092 |
| ·02 | ·01179 117 | 1 302 | ·02699 329 | 113 305 | ·52 | ·30296 722 | 34 905 | ·15777 666 | 15 146 |
| ·03 | ·02205 271 | 2 430 | ·02512 739 | 113 305 | ·53 | ·30121 522 | 34 736 | ·19259 601 | 19 197 |
| ·04 | ·03228 994 | 3 566 | ·02212 867 | 113 178 | ·54 | ·29911 594 | 34 521 | ·22722 342 | 23 232 |
| 11·05 | +0·04249 152 | − 4 697 | −1·01799 840 | +112 926 | 11·55 | +0·29667 152 | −34 269 | +0·26161 854 | − 27 253 |
| ·06 | ·05264 614 | 5 821 | ·01273 910 | 112 548 | ·56 | ·29388 448 | 33 978 | ·29574 117 | 31 250 |
| ·07 | ·06274 255 | 6 948 | 1·00635 455 | 112 045 | ·57 | ·29075 774 | 33 644 | ·32955 134 | 35 225 |
| ·08 | ·07276 950 | 8 063 | 0·99884 979 | 111 412 | ·58 | ·28729 463 | 33 274 | ·36300 932 | 39 168 |
| ·09 | ·08271 583 | 9 175 | ·99023 114 | 110 658 | ·59 | ·28349 886 | 32 861 | ·39607 569 | 43 073 |
| 11·10 | +0·09257 043 | − 10 275 | −0·98050 615 | +109 774 | 11·60 | +0·27937 455 | −32 411 | +0·42871 140 | − 46 943 |
| ·11 | ·10232 229 | 11 369 | ·96968 365 | 108 768 | ·61 | ·27492 620 | 31 923 | ·46087 777 | 50 763 |
| ·12 | ·11196 048 | 12 451 | ·95777 370 | 107 637 | ·62 | ·27015 869 | 31 397 | ·49253 660 | 54 537 |
| ·13 | ·12147 418 | 13 521 | ·94478 761 | 106 381 | ·63 | ·26507 728 | 30 832 | ·52365 016 | 58 254 |
| ·14 | ·13085 269 | 14 579 | ·93073 793 | 105 004 | ·64 | ·25968 762 | 30 230 | ·55418 128 | 61 917 |
| 11·15 | +0·14008 544 | − 15 621 | −0·91563 843 | +103 509 | 11·65 | +0·25399 572 | −29 595 | +0·58409 335 | − 65 515 |
| ·16 | ·14916 201 | 16 650 | ·89950 407 | 101 886 | ·66 | ·24800 794 | 28 921 | ·61335 040 | 69 043 |
| ·17 | ·15807 212 | 17 657 | ·88235 106 | 100 152 | ·67 | ·24173 101 | 28 213 | ·64191 715 | 72 502 |
| ·18 | ·16680 569 | 18 652 | ·86419 675 | 98 297 | ·68 | ·23517 201 | 27 471 | ·66975 902 | 75 885 |
| ·19 | ·17535 278 | 19 623 | ·84505 968 | 96 327 | ·69 | ·22833 836 | 26 696 | ·69684 219 | 79 185 |
| 11·20 | +0·18370 367 | − 20 577 | −0·82495 955 | + 94 243 | 11·70 | +0·22123 781 | −25 889 | +0·72313 366 | − 82 404 |
| ·21 | ·19184 883 | 21 508 | ·80391 719 | 92 049 | ·71 | ·21387 843 | 25 047 | ·74860 126 | 85 531 |
| ·22 | ·19977 895 | 22 419 | ·78195 454 | 89 744 | ·72 | ·20626 863 | 24 178 | ·77321 372 | 88 569 |
| ·23 | ·20748 493 | 23 302 | ·75909 465 | 87 331 | ·73 | ·19841 711 | 23 277 | ·79694 067 | 91 506 |
| ·24 | ·21495 793 | 24 164 | ·73536 164 | 84 815 | ·74 | ·19033 287 | 22 348 | ·81975 274 | 94 346 |
| 11·25 | +0·22218 934 | − 24 999 | −0·71078 067 | + 82 194 | 11·75 | +0·18202 520 | −21 391 | +0·84162 154 | − 97 082 |
| ·26 | ·22917 081 | 25 808 | ·68537 794 | 79 476 | ·76 | ·17350 367 | 20 405 | ·86251 972 | 99 707 |
| ·27 | ·23589 425 | 26 588 | ·65918 063 | 76 657 | ·77 | ·16477 813 | 19 398 | ·88242 103 | 102 225 |
| ·28 | ·24235 186 | 27 339 | ·63221 692 | 73 747 | ·78 | ·15585 866 | 18 362 | ·90130 030 | 104 627 |
| ·29 | ·24853 613 | 28 063 | ·60151 591 | 70 711 | ·79 | ·14675 561 | 17 305 | ·91913 352 | 106 910 |
| 11·30 | +0·25443 983 | − 28 756 | −0·57610 762 | + 67 654 | 11·80 | +0·13747 955 | −16 223 | +0·93589 786 | −109 074 |
| ·31 | ·26005 603 | 29 414 | ·54702 295 | 64 476 | ·81 | ·12804 129 | 15 125 | ·95157 169 | 111 112 |
| ·32 | ·26537 814 | 30 046 | ·51729 367 | 61 220 | ·82 | ·11845 182 | 14 002 | ·96613 463 | 113 027 |
| ·33 | ·27039 986 | 30 639 | ·48695 234 | 57 884 | ·83 | ·10872 236 | 12 863 | ·97956 754 | 114 808 |
| ·34 | ·27511 525 | 31 201 | ·45603 231 | 54 473 | ·84 | ·09886 430 | 11 708 | 0·99185 261 | 116 459 |
| 11·35 | +0·27951 869 | − 31 728 | −0·42456 768 | + 50 989 | 11·85 | +0·08888 919 | −10 535 | +1·00297 333 | −117 979 |
| ·36 | ·28360 491 | 32 222 | ·39259 328 | 47 441 | ·86 | ·07880 876 | 9 347 | ·01291 452 | 119 356 |
| ·37 | ·28736 898 | 32 678 | ·36014 459 | 43 829 | ·87 | ·06863 488 | 8 148 | ·02166 240 | 120 600 |
| ·38 | ·29080 634 | 33 098 | ·32725 772 | 40 155 | ·88 | ·05837 954 | 6 939 | ·02920 454 | 121 700 |
| ·39 | ·29391 279 | 33 480 | ·29396 940 | 36 428 | ·89 | ·04805 484 | 5 713 | ·03552 994 | 122 657 |
| 11·40 | +0·29668 451 | − 33 826 | −0·26031 690 | + 32 648 | 11·90 | +0·03767 302 | − 4 483 | +1·04062 903 | −123 474 |
| ·41 | ·29911 804 | 34 133 | ·22633 801 | 28 821 | ·91 | ·02724 638 | 3 246 | ·04449 365 | 124 141 |
| ·42 | ·30121 031 | 34 402 | ·19207 099 | 24 950 | ·92 | ·01678 729 | 2 002 | ·04711 713 | 124 663 |
| ·43 | ·30295 863 | 34 632 | ·15755 454 | 21 043 | ·93 | + ·00630 819 | − 752 | ·04849 425 | 125 037 |
| ·44 | ·30436 070 | 34 822 | ·12282 773 | 17 097 | ·94 | − ·00417 843 | + 499 | ·04862 127 | 125 263 |
| 11·45 | +0·30541 462 | − 34 976 | −0·08793 000 | + 13 125 | 11·95 | −0·01466 006 | + 1 752 | +1·04749 594 | −125 338 |
| ·46 | ·30611 886 | 35 084 | ·05290 107 | 9 124 | ·96 | ·02512 417 | 3 006 | ·04511 751 | 125 261 |
| ·47 | ·30647 233 | 35 157 | − ·01778 093 | 5 108 | ·97 | ·03555 823 | 4 255 | ·04148 674 | 125 039 |
| ·48 | ·30647 431 | 35 188 | + ·01739 025 | + 1 069 | ·98 | ·04594 974 | 5 506 | ·03660 587 | 124 658 |
| ·49 | ·30612 449 | 35 177 | ·05257 210 | − 2 979 | ·99 | ·05628 620 | 6 749 | ·03047 869 | 124 135 |
| 11·50 | +0·30542 297 | − 35 129 | +0·08772 415 | − 7 033 | 12·00 | −0·06655 518 | + 7 989 | +1·02311 045 | −123 452 |

TABLE I—Ai(x) AND Ai'(x)

| x | Ai(-x) | $\delta^2_m$ | Ai'(-x) | $\delta^2_m$ | x | Ai(-x) | $\delta^2_m$ | Ai'(-x) | $\delta^2_m$ |
|---|---|---|---|---|---|---|---|---|---|
| 12·00 | −0·06655 518 | + 7 989 | +1·02311 045 | − 123 452 | 12·50 | −0·27627 456 | +34 539 | −0·41933 133 | + 49 658 |
| ·01 | ·07674 429 | 9 216 | ·01450 796 | 122 626 | ·51 | ·27190 942 | 34 020 | ·45361 007 | 54 034 |
| ·02 | ·08684 125 | 10 440 | 1·00467 949 | 121 647 | ·52 | ·26720 416 | 33 458 | ·48734 858 | 58 349 |
| ·03 | ·09683 383 | 11 650 | 0·99363 483 | 120 516 | ·53 | ·26216 440 | 32 855 | ·52050 371 | 62 606 |
| ·04 | ·10670 993 | 12 850 | ·98138 528 | 119 241 | ·54 | ·25679 617 | 32 205 | ·55303 291 | 66 788 |
| 12·05 | −0·11645 756 | +14 035 | +0·96794 359 | − 117 817 | 12·55 | −0·25110 596 | +31 517 | −0·58489 436 | + 70 902 |
| ·06 | ·12606 487 | 15 203 | ·95332 400 | 116 247 | ·56 | ·24510 065 | 30 790 | ·61604 694 | 74 934 |
| ·07 | ·13552 017 | 16 361 | ·93754 221 | 114 530 | ·57 | ·23878 752 | 30 019 | ·64645 034 | 78 879 |
| ·08 | ·14481 190 | 17 494 | ·92061 538 | 112 672 | ·58 | ·23217 427 | 29 211 | ·67606 511 | 82 736 |
| ·09 | ·15392 872 | 18 613 | ·90256 209 | 110 674 | ·59 | ·22526 898 | 28 366 | ·70485 269 | 86 499 |
| 12·10 | −0·16285 945 | +19 706 | +0·88340 232 | − 108 534 | 12·60 | −0·21808 010 | +27 482 | −0·73277 547 | + 90 159 |
| ·11 | ·17159 315 | 20 784 | ·86315 746 | 106 258 | ·61 | ·21061 647 | 26 561 | ·75979 685 | 93 715 |
| ·12 | ·18011 906 | 21 832 | ·84185 027 | 103 846 | ·62 | ·20288 729 | 25 610 | ·78588 128 | 97 161 |
| ·13 | ·18842 669 | 22 859 | ·81950 486 | 101 304 | ·63 | ·19490 208 | 24 617 | ·81099 431 | 100 492 |
| ·14 | ·19650 578 | 23 858 | ·79614 665 | 98 629 | ·64 | ·18667 075 | 23 599 | ·83510 264 | 103 703 |
| 12·15 | −0·20434 634 | +24 831 | +0·77180 238 | − 95 831 | 12·65 | −0·17820 349 | +22 547 | −0·85817 417 | +106 790 |
| ·16 | ·21193 864 | 25 776 | ·74650 003 | 92 905 | ·66 | ·16951 082 | 21 462 | ·88017 804 | 109 748 |
| ·17 | ·21927 324 | 26 687 | ·72026 885 | 89 862 | ·67 | ·16060 358 | 20 351 | ·90108 467 | 112 575 |
| ·18 | ·22634 102 | 27 572 | ·69313 927 | 86 700 | ·68 | ·15149 288 | 19 212 | ·92086 580 | 115 266 |
| ·19 | ·23313 314 | 28 422 | ·66514 290 | 83 422 | ·69 | ·14219 011 | 18 047 | ·93949 453 | 117 813 |
| 12·20 | −0·23964 110 | +29 241 | +0·63631 251 | − 80 038 | 12·70 | −0·13270 692 | +16 856 | −0·95694 539 | +120 220 |
| ·21 | ·24585 672 | 30 022 | ·60668 194 | 76 545 | ·71 | ·12305 521 | 15 643 | ·97319 432 | 122 478 |
| ·22 | ·25177 218 | 30 770 | ·57628 611 | 72 949 | ·72 | ·11324 711 | 14 405 | 0·98821 875 | 124 584 |
| ·23 | ·25738 001 | 31 481 | ·54516 097 | 69 258 | ·73 | ·10329 499 | 13 153 | 1·00199 762 | 126 537 |
| ·24 | ·26267 310 | 32 157 | ·51334 343 | 65 468 | ·74 | ·09321 138 | 11 877 | ·01451 141 | 128 334 |
| 12·25 | −0·26764 470 | +32 790 | +0·48087 137 | − 61 592 | 12·75 | −0·08300 903 | +10 583 | −1·02574 216 | +129 968 |
| ·26 | ·27228 847 | 33 386 | ·44778 355 | 57 630 | ·76 | ·07270 087 | 9 281 | ·03567 353 | 131 441 |
| ·27 | ·27659 845 | 33 942 | ·41411 958 | 53 586 | ·77 | ·06229 994 | 7 954 | ·04429 079 | 132 751 |
| ·28 | ·28056 908 | 34 459 | ·37991 989 | 49 467 | ·78 | ·05181 948 | 6 626 | ·05158 085 | 133 890 |
| ·29 | ·28419 520 | 34 932 | ·34522 566 | 45 276 | ·79 | ·04127 279 | 5 279 | ·05753 232 | 134 862 |
| 12·30 | −0·28747 208 | +35 363 | +0·31007 879 | − 41 022 | 12·80 | −0·03067 332 | + 3 926 | −1·06213 548 | +135 665 |
| ·31 | ·29039 541 | 35 754 | ·27452 182 | 36 703 | ·81 | ·02003 460 | 2 567 | ·06538 231 | 136 291 |
| ·32 | ·29296 129 | 36 095 | ·23859 792 | 32 329 | ·82 | − ·00937 022 | + 1 202 | ·06726 654 | 136 749 |
| ·33 | ·29516 629 | 36 400 | ·20235 082 | 27 906 | ·83 | + ·00130 617 | − 166 | ·06778 361 | 137 027 |
| ·34 | ·29700 738 | 36 655 | ·16582 474 | 23 438 | ·84 | ·01198 089 | 1 539 | ·06693 073 | 137 130 |
| 12·35 | −0·29848 200 | +36 866 | +0·12906 436 | − 18 927 | 12·85 | +0·02264 022 | − 2 909 | −1·06470 687 | +137 061 |
| ·36 | ·29958 804 | 37 035 | ·09211 477 | 14 385 | ·86 | ·03327 046 | 4 280 | ·06111 273 | 136 808 |
| ·37 | ·30032 382 | 37 154 | ·05502 139 | 9 811 | ·87 | ·04385 791 | 5 643 | ·05615 083 | 136 384 |
| ·38 | ·30068 814 | 37 231 | + ·01782 994 | 5 216 | ·88 | ·05438 893 | 7 008 | ·04982 542 | 135 778 |
| ·39 | ·30068 024 | 37 258 | − ·01941 364 | − 604 | ·89 | ·06484 989 | 8 358 | ·04214 255 | 134 999 |
| 12·40 | −0·30029 984 | +37 241 | −0·05666 323 | + 4 025 | 12·90 | +0·07522 728 | − 9 706 | −1·03311 002 | +134 041 |
| ·41 | ·29954 711 | 37 179 | ·09387 257 | 8 652 | ·91 | ·08550 763 | 11 040 | ·02273 740 | 132 908 |
| ·42 | ·29842 268 | 37 068 | ·13099 538 | 13 286 | ·92 | ·09567 760 | 12 364 | 1·01103 602 | 131 600 |
| ·43 | ·29692 765 | 36 914 | ·16798 534 | 17 913 | ·93 | ·10572 396 | 13 670 | 0·99801 896 | 130 119 |
| ·44 | ·29506 357 | 36 710 | ·20479 619 | 22 527 | ·94 | ·11563 364 | 14 965 | ·98370 103 | 128 464 |
| 12·45 | −0·29283 247 | +36 461 | −0·24138 180 | + 27 126 | 12·95 | +0·12539 370 | −16 241 | −0·96809 877 | +126 640 |
| ·46 | ·29023 684 | 36 169 | ·27769 619 | 31 703 | ·96 | ·13499 139 | 17 495 | ·95123 042 | 124 646 |
| ·47 | ·28727 961 | 35 828 | ·31369 361 | 36 247 | ·97 | ·14441 416 | 18 735 | ·93311 592 | 122 485 |
| ·48 | ·28396 418 | 35 445 | ·34932 862 | 40 760 | ·98 | ·15364 963 | 19 945 | ·91377 687 | 120 162 |
| ·49 | ·28029 439 | 35 012 | ·38455 610 | 45 234 | ·99 | ·16268 569 | 21 136 | ·89323 650 | 117 674 |
| 12·50 | −0·27627 456 | +34 539 | −0·41933 133 | + 49 658 | 13·00 | +0·17151 044 | −22 299 | −0·87151 968 | +115 029 |

B 32

TABLE I—Ai($x$) AND Ai$'(x)$

| $x$ | Ai($-x$) | $\delta^2_m$ | Ai$'(-x)$ | $\delta^2_m$ | $x$ | Ai($-x$) | $\delta^2_m$ | Ai$'(-x)$ | $\delta^2_m$ |
|---|---|---|---|---|---|---|---|---|---|
| 13·00 | +0·17151 044 | − 22 299 | −0·87151 968 | + 115 029 | 13·50 | +0·19098 124 | − 25 784 | +0·82643 275 | − 109 671 |
| ·01 | ·18011 225 | 23 435 | ·84865 286 | 112 224 | ·51 | ·18258 985 | 24 673 | ·85166 091 | 113 249 |
| ·02 | ·18847 976 | 24 544 | ·82466 407 | 109 274 | ·52 | ·17395 180 | 23 521 | ·87575 685 | 116 678 |
| ·03 | ·19660 189 | 25 619 | ·79958 283 | 106 164 | ·53 | ·16507 860 | 22 339 | ·89868 629 | 119 956 |
| ·04 | ·20446 788 | 26 668 | ·77344 021 | 102 916 | ·54 | ·15598 207 | 21 122 | ·92041 645 | 123 082 |
| 13·05 | +0·21206 726 | − 27 677 | −0·74626 869 | + 99 524 | 13·55 | +0·14667 437 | − 19 877 | +0·94091 609 | − 126 043 |
| ·06 | ·21938 993 | 28 656 | ·71810 219 | 95 990 | ·56 | ·13716 795 | 18 603 | ·96015 560 | 128 844 |
| ·07 | ·22642 611 | 29 598 | ·68897 603 | 92 327 | ·57 | ·12747 555 | 17 301 | ·97810 699 | 131 470 |
| ·08 | ·23316 638 | 30 504 | ·65892 684 | 88 532 | ·58 | ·11761 019 | 15 973 | 0·99474 399 | 133 929 |
| ·09 | ·23960 169 | 31 365 | ·62799 256 | 84 612 | ·59 | ·10758 514 | 14 625 | 1·01004 203 | 136 207 |
| 13·10 | +0·24572 341 | − 32 195 | −0·59621 238 | + 80 572 | 13·60 | +0·09741 389 | − 13 249 | +1·02397 833 | − 138 304 |
| ·11 | ·25152 326 | 32 980 | ·56362 669 | 76 418 | ·61 | ·08711 018 | 11 857 | ·03653 192 | 140 221 |
| ·12 | ·25699 339 | 33 721 | ·53027 702 | 72 154 | ·62 | ·07668 793 | 10 447 | ·04768 365 | 141 946 |
| ·13 | ·26212 639 | 34 421 | ·49620 601 | 67 782 | ·63 | ·06616 124 | 9 020 | ·05741 627 | 143 483 |
| ·14 | ·26691 526 | 35 079 | ·46145 736 | 63 314 | ·64 | ·05554 438 | 7 577 | ·06571 441 | 144 830 |
| 13·15 | +0·27135 343 | − 35 687 | −0·42607 574 | + 58 753 | 13·65 | +0·04485 177 | − 6 123 | +1·07256 462 | − 145 975 |
| ·16 | ·27543 481 | 36 252 | ·39010 676 | 54 100 | ·66 | ·03409 795 | 4 659 | ·07795 544 | 146 928 |
| ·17 | ·27915 376 | 36 768 | ·35359 693 | 49 368 | ·67 | ·02329 756 | 3 185 | ·08187 735 | 147 682 |
| ·18 | ·28250 511 | 37 240 | ·31659 356 | 44 559 | ·68 | ·01246 533 | 1 706 | ·08432 282 | 148 230 |
| ·19 | ·28548 415 | 37 660 | ·27914 473 | 39 681 | ·69 | + ·00161 605 | − 223 | ·08528 636 | 148 581 |
| 13·20 | +0·28808 668 | − 38 034 | −0·24129 921 | + 34 737 | 13·70 | −0·00923 545 | + 1 266 | +1·08476 447 | − 148 726 |
| ·21 | ·29030 897 | 38 352 | ·20310 642 | 29 740 | ·71 | ·02007 429 | 2 752 | ·08275 570 | 148 666 |
| ·22 | ·29214 782 | 38 629 | ·16461 633 | 24 686 | ·72 | ·03088 561 | 4 239 | ·07926 064 | 148 405 |
| ·23 | ·29360 048 | 38 848 | ·12587 945 | 19 596 | ·73 | ·04165 455 | 5 719 | ·07428 191 | 147 935 |
| ·24 | ·29466 475 | 39 018 | ·08694 669 | 14 460 | ·74 | ·05236 631 | 7 195 | ·06782 420 | 147 264 |
| 13·25 | +0·29533 893 | − 39 138 | −0·04786 938 | + 9 298 | 13·75 | −0·06300 613 | + 8 665 | +1·05989 423 | − 146 385 |
| ·26 | ·29562 183 | 39 204 | − ·00869 913 | + 4 112 | ·76 | ·07355 932 | 10 123 | ·05050 078 | 145 306 |
| ·27 | ·29551 278 | 39 221 | + ·03051 221 | − 1 093 | ·77 | ·08401 130 | 11 570 | ·03965 465 | 144 021 |
| ·28 | ·29501 162 | 39 183 | ·06971 260 | 6 307 | ·78 | ·09434 761 | 13 001 | ·02736 868 | 142 535 |
| ·29 | ·29411 873 | 39 092 | ·10884 991 | 11 526 | ·79 | ·10455 393 | 14 422 | 1·01365 773 | 140 848 |
| 13·30 | +0·29283 501 | − 38 954 | +0·14787 197 | − 16 738 | 13·80 | −0·11461 607 | + 15 817 | +0·99853 866 | − 138 966 |
| ·31 | ·29116 185 | 38 757 | ·18672 666 | 21 945 | ·81 | ·12452 007 | 17 199 | ·98203 030 | 136 881 |
| ·32 | ·28910 121 | 38 515 | ·22536 194 | 27 129 | ·82 | ·13425 212 | 18 556 | ·96415 348 | 134 610 |
| ·33 | ·28665 552 | 38 216 | ·26372 597 | 32 291 | ·83 | ·14379 865 | 19 890 | ·94493 092 | 132 139 |
| ·34 | ·28382 776 | 37 867 | ·30176 715 | 37 423 | ·84 | ·15314 633 | 21 198 | ·92438 731 | 129 487 |
| 13·35 | +0·28062 142 | − 37 467 | +0·33943 418 | − 42 512 | 13·85 | −0·16228 208 | + 22 478 | +0·90254 918 | − 126 645 |
| ·36 | ·27704 050 | 37 020 | ·37667 617 | 47 559 | ·86 | ·17119 310 | 23 732 | ·87944 494 | 123 619 |
| ·37 | ·27308 948 | 36 516 | ·41344 267 | 52 553 | ·87 | ·17986 686 | 24 950 | ·85510 483 | 120 421 |
| ·38 | ·26877 339 | 35 966 | ·44968 375 | 57 486 | ·88 | ·18829 118 | 26 138 | ·82956 084 | 117 043 |
| ·39 | ·26409 773 | 35 370 | ·48535 009 | 62 354 | ·89 | ·19645 418 | 27 292 | ·80284 673 | 113 495 |
| 13·40 | +0·25906 847 | − 34 719 | +0·52039 302 | − 67 151 | 13·90 | −0·20434 433 | + 28 407 | +0·77499 797 | − 109 787 |
| ·41 | ·25369 210 | 34 025 | ·55476 459 | 71 866 | ·91 | ·21195 048 | 29 486 | ·74605 165 | 105 909 |
| ·42 | ·24797 557 | 33 282 | ·58841 766 | 76 493 | ·92 | ·21926 184 | 30 527 | ·71604 652 | 101 884 |
| ·43 | ·24192 630 | 32 497 | ·62130 596 | 81 034 | ·93 | ·22626 801 | 31 521 | ·68502 284 | 97 701 |
| ·44 | ·23555 215 | 31 660 | ·65338 411 | 85 469 | ·94 | ·23295 904 | 32 481 | ·65302 242 | 93 373 |
| 13·45 | +0·22886 147 | − 30 789 | +0·68460 776 | − 89 803 | 13·95 | −0·23932 535 | + 33 390 | +0·62008 852 | − 88 912 |
| ·46 | ·22186 299 | 29 865 | ·71493 359 | 94 024 | ·96 | ·24535 784 | 34 255 | ·58626 576 | 84 308 |
| ·47 | ·21456 593 | 28 908 | ·74431 940 | 98 125 | ·97 | ·25104 786 | 35 079 | ·55160 015 | 79 582 |
| ·48 | ·20697 987 | 27 902 | ·77272 418 | 102 107 | ·98 | ·25638 719 | 35 846 | ·51613 895 | 74 731 |
| ·49 | ·19911 485 | 26 867 | ·80010 813 | 105 958 | ·99 | ·26136 814 | 36 570 | ·47993 065 | 69 767 |
| 13·50 | +0·19098 124 | − 25 784 | +0·82643 275 | − 109 671 | 14·00 | −0·26598 348 | + 37 242 | +0·44302 488 | − 64 694 |

## TABLE I—Ai($x$) AND Ai'($x$)

| $x$ | Ai($-x$) | $\delta_m^2$ | Ai'($-x$) | $\delta_m^2$ | $x$ | Ai($-x$) | $\delta_m^2$ | Ai'($-x$) | $\delta_m^2$ |
|---|---|---|---|---|---|---|---|---|---|
| 14·00 | −0·26598 348 | + 37 242 | +0·44302 488 | − 64 694 | 14·50 | −0·03059 742 | + 4 438 | −1·09532 127 | +158 538 |
| ·01 | ·27022 649 | 37 865 | ·40547 236 | 59 520 | ·51 | ·01962 467 | 2 848 | ·09896 398 | 159 288 |
| ·02 | ·27409 095 | 38 433 | ·36732 482 | 54 246 | ·52 | − ·00862 345 | + 1 253 | ·10101 424 | 159 803 |
| ·03 | ·27757 118 | 38 949 | ·32863 497 | 48 894 | ·53 | + ·00239 029 | − 348 | ·10146 689 | 160 091 |
| ·04 | ·28066 202 | 39 411 | ·28945 634 | 43 452 | ·54 | ·01340 055 | 1 948 | ·10031 906 | 160 144 |
| 14·05 | −0·28335 885 | + 39 815 | +0·24984 332 | − 37 945 | 14·55 | +0·02439 133 | − 3 548 | −1·09757 022 | +159 964 |
| ·06 | ·28565 762 | 40 171 | ·20985 098 | 32 366 | ·56 | ·03534 663 | 5 148 | ·09322 217 | 159 550 |
| ·07 | ·28755 479 | 40 463 | ·16953 508 | 26 734 | ·57 | ·04625 046 | 6 739 | ·08727 905 | 158 903 |
| ·08 | ·28904 743 | 40 705 | ·12895 193 | 21 052 | ·58 | ·05708 691 | 8 324 | ·07974 733 | 158 021 |
| ·09 | ·29013 313 | 40 884 | ·08815 834 | 15 326 | ·59 | ·06784 014 | 9 899 | ·07063 583 | 156 908 |
| 14·10 | −0·29081 009 | + 41 013 | +0·04721 155 | − 9 568 | 14·60 | +0·07849 440 | − 11 462 | −1·05995 568 | +155 563 |
| ·11 | ·29107 704 | 41 075 | + ·00616 913 | − 3 782 | ·61 | ·08903 407 | 13 009 | ·04772 033 | 153 983 |
| ·12 | ·29093 334 | 41 086 | − ·03491 108 | + 2 018 | ·62 | ·09944 368 | 14 542 | ·03394 556 | 152 181 |
| ·13 | ·29037 889 | 41 037 | ·07597 109 | 7 832 | ·63 | ·10970 791 | 16 052 | ·01864 940 | 150 148 |
| ·14 | ·28941 418 | 40 928 | ·11695 278 | 13 643 | ·64 | ·11981 166 | 17 542 | 1·00185 217 | 147 890 |
| 14·15 | −0·28804 029 | + 40 764 | −0·15779 805 | + 19 448 | 14·65 | +0·12974 003 | − 19 012 | −0·98357 644 | +145 415 |
| ·16 | ·28625 887 | 40 541 | ·19844 886 | 25 242 | ·66 | ·13947 834 | 20 448 | ·96384 697 | 142 715 |
| ·17 | ·28407 215 | 40 257 | ·23884 730 | 31 007 | ·67 | ·14901 221 | 21 864 | ·94269 074 | 139 805 |
| ·18 | ·28148 296 | 39 922 | ·27893 573 | 36 741 | ·68 | ·15832 750 | 23 246 | ·92013 685 | 136 680 |
| ·19 | ·27849 466 | 39 523 | ·31865 682 | 42 440 | ·69 | ·16741 039 | 24 595 | ·89621 654 | 133 348 |
| 14·20 | −0·27511 123 | + 39 071 | −0·35795 361 | + 48 081 | 14·70 | +0·17624 739 | − 25 913 | −0·87096 312 | +129 815 |
| ·21 | ·27133 719 | 38 562 | ·39676 968 | 53 676 | ·71 | ·18482 533 | 27 192 | ·84441 192 | 126 079 |
| ·22 | ·26717 763 | 38 000 | ·43504 912 | 59 199 | ·72 | ·19313 142 | 28 432 | ·81660 028 | 122 155 |
| ·23 | ·26263 818 | 37 377 | ·47273 670 | 64 653 | ·73 | ·20115 326 | 29 634 | ·78756 744 | 118 038 |
| ·24 | ·25772 505 | 36 706 | ·50977 790 | 70 023 | ·74 | ·20887 884 | 30 795 | ·75735 455 | 113 740 |
| 14·25 | −0·25244 496 | + 35 979 | −0·54611 903 | + 75 306 | 14·75 | +0·21629 656 | − 31 907 | −0·72600 458 | +109 267 |
| ·26 | ·24680 518 | 35 199 | ·58170 727 | 80 496 | ·76 | ·22339 529 | 32 979 | ·69356 226 | 104 619 |
| ·27 | ·24081 350 | 34 369 | ·61649 075 | 85 575 | ·77 | ·23016 432 | 34 000 | ·66007 405 | 99 811 |
| ·28 | ·23447 822 | 33 489 | ·65041 868 | 90 547 | ·78 | ·23659 344 | 34 973 | ·62558 802 | 94 842 |
| ·29 | ·22780 814 | 32 557 | ·68344 136 | 95 398 | ·79 | ·24267 292 | 35 897 | ·59015 384 | 89 726 |
| 14·30 | −0·22081 257 | + 31 582 | −0·71551 029 | +100 124 | 14·80 | +0·24839 353 | − 36 768 | −0·55382 267 | + 84 463 |
| ·31 | ·21350 127 | 30 556 | ·74657 823 | 104 715 | ·81 | ·25374 656 | 37 585 | ·51664 712 | 79 067 |
| ·32 | ·20588 449 | 29 487 | ·77659 928 | 109 164 | ·82 | ·25872 384 | 38 350 | ·47868 114 | 73 539 |
| ·33 | ·19797 292 | 28 372 | ·80552 896 | 113 469 | ·83 | ·26331 773 | 39 055 | ·43997 999 | 67 894 |
| ·34 | ·18977 770 | 27 220 | ·83332 424 | 117 618 | ·84 | ·26752 117 | 39 706 | ·40060 011 | 62 134 |
| 14·35 | −0·18131 036 | + 26 020 | −0·85994 364 | +121 606 | 14·85 | +0·27132 766 | − 40 298 | −0·36059 908 | + 56 273 |
| ·36 | ·17258 288 | 24 789 | ·88534 729 | 125 426 | ·86 | ·27473 128 | 40 831 | ·32003 550 | 50 313 |
| ·37 | ·16360 759 | 23 512 | ·90949 699 | 129 077 | ·87 | ·27772 670 | 41 306 | ·27896 895 | 44 269 |
| ·38 | ·15439 724 | 22 207 | ·93235 625 | 132 548 | ·88 | ·28030 918 | 41 715 | ·23745 986 | 38 144 |
| ·39 | ·14496 489 | 20 864 | ·95389 037 | 135 834 | ·89 | ·28247 462 | 42 066 | ·19556 946 | 31 951 |
| 14·40 | −0·13532 396 | + 19 489 | −0·97406 650 | +138 932 | 14·90 | +0·28421 951 | − 42 356 | −0·15335 966 | + 25 697 |
| ·41 | ·12548 819 | 18 087 | 0·99285 367 | 141 836 | ·91 | ·28554 096 | 42 581 | ·11089 298 | 19 395 |
| ·42 | ·11547 161 | 16 652 | 1·01022 285 | 144 541 | ·92 | ·28643 672 | 42 743 | ·06823 243 | 13 046 |
| ·43 | ·10528 855 | 15 197 | ·02614 700 | 147 040 | ·93 | ·28690 517 | 42 840 | − ·02544 147 | 6 670 |
| ·44 | ·09495 357 | 13 713 | ·04060 113 | 149 335 | ·94 | ·28694 533 | 42 877 | + ·01741 615 | + 270 |
| 14·45 | −0·08448 150 | + 12 210 | −1·05356 230 | +151 417 | 14·95 | +0·28655 684 | − 42 847 | +0·06027 644 | − 6 147 |
| ·46 | ·07388 737 | 10 686 | ·06500 970 | 153 283 | ·96 | ·28574 000 | 42 753 | ·10307 526 | 12 563 |
| ·47 | ·06318 641 | 9 143 | ·07492 467 | 154 931 | ·97 | ·28449 575 | 42 594 | ·14574 846 | 18 975 |
| ·48 | ·05239 404 | 7 589 | ·08329 073 | 156 361 | ·98 | ·28282 567 | 42 375 | ·18823 194 | 25 372 |
| ·49 | ·04152 581 | 6 018 | ·09009 360 | 157 561 | ·99 | ·28073 196 | 42 087 | ·23046 175 | 31 742 |
| 14·50 | −0·03059 742 | + 4 438 | −1·09532 127 | +158 538 | 15·00 | +0·27821 749 | − 41 738 | +0·27237 420 | − 38 078 |

B 34

# TABLE I—Ai(x) AND Ai'(x)

| x | Ai(−x) | $\delta^2_m$ | Ai'(−x) | $\delta^2_m$ | x | Ai(−x) | $\delta^2_m$ | Ai'(−x) | $\delta^2_m$ |
|---|---|---|---|---|---|---|---|---|---|
| 15·00 | +0·27821 749 | − 41 738 | +0·27237 420 | − 38 078 | 15·50 | −0·16644 795 | + 25 801 | +0·90493 794 | − 141 955 |
| ·01 | ·27528 575 | 41 328 | ·31390 595 | 44 370 | ·51 | ·17536 599 | 27 205 | ·87843 531 | 138 020 |
| ·02 | ·27194 085 | 40 851 | ·35499 410 | 50 606 | ·52 | ·18401 206 | 28 563 | ·85055 288 | 133 868 |
| ·03 | ·26818 755 | 40 314 | ·39557 630 | 56 782 | ·53 | ·19237 258 | 29 879 | ·82133 217 | 129 498 |
| ·04 | ·26403 122 | 39 718 | ·43559 082 | 62 880 | ·54 | ·20043 439 | 31 153 | ·79081 687 | 124 917 |
| 15·05 | +0·25947 783 | − 39 057 | +0·47497 669 | − 68 898 | 15·55 | −0·20818 476 | + 32 378 | +0·75905 277 | − 120 134 |
| ·06 | ·25453 398 | 38 339 | ·51367 375 | 74 827 | ·56 | ·21561 144 | 33 554 | ·72608 769 | 115 155 |
| ·07 | ·24920 685 | 37 562 | ·55162 274 | 80 646 | ·57 | ·22270 267 | 34 680 | ·69197 141 | 109 985 |
| ·08 | ·24350 421 | 36 725 | ·58876 546 | 86 365 | ·58 | ·22944 720 | 35 753 | ·65675 561 | 104 635 |
| ·09 | ·23743 442 | 35 836 | ·62504 476 | 91 956 | ·59 | ·23583 430 | 36 772 | ·62049 378 | 99 109 |
| 15·10 | +0·23100 638 | − 34 887 | +0·66040 473 | − 97 425 | 15·60 | −0·24185 378 | + 37 736 | +0·58324 116 | − 93 420 |
| ·11 | ·22422 957 | 33 887 | ·69479 070 | 102 756 | ·61 | ·24749 601 | 38 640 | ·54505 463 | 87 572 |
| ·12 | ·21711 399 | 32 832 | ·72814 938 | 107 940 | ·62 | ·25275 195 | 39 486 | ·50599 265 | 81 579 |
| ·13 | ·20967 018 | 31 728 | ·76042 894 | 112 973 | ·63 | ·25761 314 | 40 270 | ·46611 514 | 75 441 |
| ·14 | ·20190 918 | 30 574 | ·79157 907 | 117 843 | ·64 | ·26207 174 | 40 995 | ·42548 345 | 69 180 |
| 15·15 | +0·19384 253 | − 29 373 | +0·82155 108 | − 122 543 | 15·65 | −0·26612 051 | + 41 655 | +0·38416 019 | − 62 793 |
| ·16 | ·18548 224 | 28 123 | ·85029 798 | 127 071 | ·66 | ·26975 285 | 42 248 | ·34220 920 | 56 299 |
| ·17 | ·17684 080 | 26 831 | ·87777 452 | 131 407 | ·67 | ·27296 282 | 42 782 | ·29969 541 | 49 699 |
| ·18 | ·16793 113 | 25 497 | ·90393 733 | 135 560 | ·68 | ·27574 510 | 43 242 | ·25668 479 | 43 014 |
| ·19 | ·15876 657 | 24 120 | ·92874 491 | 139 508 | ·69 | ·27809 508 | 43 641 | ·21324 418 | 36 247 |
| 15·20 | +0·14936 088 | − 22 707 | +0·95215 778 | − 143 256 | 15·70 | −0·28000 878 | + 43 969 | +0·16944 123 | − 29 408 |
| ·21 | ·13972 819 | 21 256 | ·97413 848 | 146 789 | ·71 | ·28148 292 | 44 228 | ·12534 431 | 22 511 |
| ·22 | ·12988 300 | 19 771 | 0·99465 168 | 150 112 | ·72 | ·28251 491 | 44 418 | ·08102 237 | 15 566 |
| ·23 | ·11984 016 | 18 256 | 1·01366 418 | 153 204 | ·73 | ·28310 285 | 44 539 | + ·03654 484 | 8 583 |
| ·24 | ·10961 482 | 16 706 | ·03114 505 | 156 074 | ·74 | ·28324 553 | 44 591 | − ·00801 847 | − 1 570 |
| 15·25 | +0·09922 246 | − 15 135 | +1·04706 561 | − 158 710 | 15·75 | −0·28294 243 | + 44 570 | −0·05259 746 | + 5 454 |
| ·26 | ·08867 880 | 13 534 | ·06139 951 | 161 108 | ·76 | ·28219 376 | 44 481 | ·09712 190 | 12 484 |
| ·27 | ·07799 984 | 11 913 | ·07412 278 | 163 262 | ·77 | ·28100 041 | 44 321 | ·14152 151 | 19 512 |
| ·28 | ·06720 179 | 10 270 | ·08521 388 | 165 174 | ·78 | ·27936 398 | 44 091 | ·18572 604 | 26 514 |
| ·29 | ·05630 107 | 8 611 | ·09465 370 | 166 836 | ·79 | ·27728 677 | 43 791 | ·22966 547 | 33 497 |
| 15·30 | +0·04531 427 | − 6 933 | +1·10242 563 | − 168 244 | 15·80 | −0·27477 178 | + 43 419 | −0·27327 001 | + 40 433 |
| ·31 | ·03425 816 | 5 246 | ·10851 559 | 169 396 | ·81 | ·27182 272 | 42 983 | ·31647 031 | 47 323 |
| ·32 | ·02314 961 | 3 549 | ·11291 206 | 170 293 | ·82 | ·26844 396 | 42 476 | ·35919 750 | 54 149 |
| ·33 | ·01200 559 | 1 839 | ·11560 608 | 170 929 | ·83 | ·26464 057 | 41 898 | ·40138 334 | 60 900 |
| ·34 | + ·00084 318 | − 129 | ·11659 129 | 171 302 | ·84 | ·26041 832 | 41 257 | ·44296 033 | 67 571 |
| 15·35 | −0·01032 052 | + 1 582 | +1·11586 396 | − 171 416 | 15·85 | −0·25578 362 | + 40 549 | −0·48386 179 | + 74 144 |
| ·36 | ·02146 839 | 3 299 | ·11342 296 | 171 262 | ·86 | ·25074 355 | 39 774 | ·52402 200 | 80 616 |
| ·37 | ·03258 328 | 5 009 | ·10926 982 | 170 848 | ·87 | ·24530 586 | 38 935 | ·56337 627 | 86 965 |
| ·38 | ·04364 809 | 6 713 | ·10340 869 | 170 168 | ·88 | ·23947 893 | 38 037 | ·60186 111 | 93 196 |
| ·39 | ·05464 578 | 8 411 | ·09584 637 | 169 223 | ·89 | ·23327 175 | 37 073 | ·63941 425 | 99 285 |
| 15·40 | −0·06555 938 | + 10 099 | +1·08659 230 | − 168 017 | 15·90 | −0·22669 395 | + 36 049 | −0·67597 481 | + 105 227 |
| ·41 | ·07637 202 | 11 768 | ·07565 854 | 166 549 | ·91 | ·21975 576 | 34 970 | ·71148 338 | 111 018 |
| ·42 | ·08706 700 | 13 429 | ·06305 977 | 164 820 | ·92 | ·21246 798 | 33 830 | ·74588 208 | 116 635 |
| ·43 | ·09762 773 | 15 065 | ·04881 327 | 162 834 | ·93 | ·20484 200 | 32 635 | ·77911 474 | 122 085 |
| ·44 | ·10803 784 | 16 685 | ·03293 890 | 160 592 | ·94 | ·19688 976 | 31 391 | ·81112 690 | 127 343 |
| 15·45 | −0·11828 115 | + 18 275 | +1·01545 908 | − 158 096 | 15·95 | −0·18862 371 | + 30 090 | −0·84186 598 | + 132 412 |
| ·46 | ·12834 175 | 19 846 | 0·99639 876 | 155 353 | ·96 | ·18005 685 | 28 742 | ·87128 131 | 137 278 |
| ·47 | ·13820 395 | 21 382 | ·97578 537 | 152 359 | ·97 | ·17120 266 | 27 347 | ·89932 425 | 141 931 |
| ·48 | ·14785 238 | 22 891 | ·95364 883 | 149 128 | ·98 | ·16207 509 | 25 902 | ·92594 828 | 146 369 |
| ·49 | ·15727 196 | 24 366 | ·93002 145 | 145 655 | ·99 | ·15268 857 | 24 420 | ·95110 904 | 150 579 |
| 15·50 | −0·16644 795 | + 25 801 | +0·90493 794 | − 141 955 | 16·00 | −0·14305 793 | + 22 893 | −0·97476 444 | + 154 557 |

# TABLE I—Ai(x) AND Ai'(x)

| x | Ai(−x) | $\delta^2_m$ | Ai'(−x) | $\delta^2_m$ | x | Ai(−x) | $\delta^2_m$ | Ai'(−x) | $\delta^2_m$ |
|---|---|---|---|---|---|---|---|---|---|
| 16·00 | −0·14305 793 | + 22 893 | −0·97476 444 | +154 557 | 16·50 | +0·27886 848 | − 46 021 | −0·09462 258 | + 18 406 |
| ·01 | ·13319 843 | 21 329 | 0·99687 472 | 158 291 | ·51 | ·27958 436 | 46 165 | ·04852 991 | 10 811 |
| ·02 | ·12312 571 | 19 728 | 1·01740 254 | 161 784 | ·52 | ·27983 872 | 46 239 | − ·00232 919 | + 3 186 |
| ·03 | ·11285 577 | 18 094 | ·03631 299 | 165 017 | ·53 | ·27963 084 | 46 230 | + ·04390 335 | − 4 461 |
| ·04 | ·10240 495 | 16 428 | ·05357 374 | 167 996 | ·54 | ·27896 080 | 46 148 | ·09009 127 | 12 112 |
| 16·05 | −0·09178 990 | + 14 736 | −1·06915 502 | +170 710 | 16·55 | +0·27782 942 | − 45 988 | +0·13615 808 | − 19 757 |
| ·06 | ·08102 754 | 13 015 | ·08302 970 | 173 149 | ·56 | ·27623 830 | 45 753 | ·18202 735 | 27 383 |
| ·07 | ·07013 507 | 11 271 | ·09517 338 | 175 323 | ·57 | ·27418 979 | 45 441 | ·22762 284 | 34 982 |
| ·08 | ·05912 992 | 9 512 | ·10556 435 | 177 212 | ·58 | ·27168 701 | 45 054 | ·27286 860 | 42 529 |
| ·09 | ·04802 969 | 7 728 | ·11418 372 | 178 819 | ·59 | ·26873 383 | 44 589 | ·31768 917 | 50 024 |
| 16·10 | −0·03685 220 | + 5 934 | −1·12101 542 | +180 146 | 16·60 | +0·26533 489 | − 44 054 | +0·36200 963 | − 57 451 |
| ·11 | ·02561 539 | 4 129 | ·12604 620 | 181 178 | ·61 | ·26149 555 | 43 441 | ·40575 574 | 64 789 |
| ·12 | ·01433 731 | 2 311 | ·12926 573 | 181 924 | ·62 | ·25722 193 | 42 757 | ·44885 413 | 72 039 |
| ·13 | − ·00303 613 | + 489 | ·13066 656 | 182 376 | ·63 | ·25252 087 | 42 001 | ·49123 233 | 79 177 |
| ·14 | + ·00826 994 | − 1 333 | ·13024 417 | 182 535 | ·64 | ·24739 993 | 41 175 | ·53281 897 | 86 202 |
| 16·15 | +0·01956 267 | − 3 161 | −1·12799 698 | +182 395 | 16·65 | +0·24186 737 | − 40 279 | +0·57354 384 | − 93 090 |
| ·16 | ·03082 380 | 4 981 | ·12392 638 | 181 966 | ·66 | ·23593 215 | 39 312 | ·61333 807 | 99 839 |
| ·17 | ·04203 513 | 6 799 | ·11803 667 | 181 238 | ·67 | ·22960 393 | 38 283 | ·65213 420 | 106 432 |
| ·18 | ·05317 849 | 8 603 | ·11033 513 | 180 211 | ·68 | ·22289 301 | 37 183 | ·68986 632 | 112 859 |
| ·19 | ·06423 583 | 10 404 | ·10083 201 | 178 899 | ·69 | ·21581 037 | 36 027 | ·72647 018 | 119 109 |
| 16·20 | +0·07518 917 | − 12 180 | −1·08954 045 | +177 286 | 16·70 | +0·20836 758 | − 34 803 | +0·76188 330 | −125 170 |
| ·21 | ·08602 073 | 13 947 | ·07647 656 | 175 385 | ·71 | ·20057 687 | 33 521 | ·79604 508 | 131 036 |
| ·22 | ·09671 286 | 15 689 | ·06165 935 | 173 198 | ·72 | ·19245 105 | 32 185 | ·82889 689 | 136 688 |
| ·23 | ·10724 814 | 17 409 | ·04511 069 | 170 723 | ·73 | ·18400 349 | 30 788 | ·86038 222 | 142 125 |
| ·24 | ·11760 938 | 19 102 | ·02685 532 | 167 965 | ·74 | ·17524 814 | 29 343 | ·89044 672 | 147 331 |
| 16·25 | +0·12777 965 | − 20 768 | −1·00692 081 | +164 929 | 16·75 | +0·16619 946 | − 27 841 | +0·91903 834 | −152 303 |
| ·26 | ·13774 230 | 22 399 | 0·98533 751 | 161 621 | ·76 | ·15687 245 | 26 298 | ·94610 739 | 157 025 |
| ·27 | ·14748 102 | 24 000 | ·96213 850 | 158 042 | ·77 | ·14728 255 | 24 704 | ·97160 666 | 161 491 |
| ·28 | ·15697 981 | 25 560 | ·93735 956 | 154 199 | ·78 | ·13744 569 | 23 066 | 0·99549 150 | 165 697 |
| ·29 | ·16622 307 | 27 082 | ·91103 911 | 150 094 | ·79 | ·12737 824 | 21 391 | 1·01771 987 | 169 627 |
| 16·30 | +0·17519 559 | − 28 563 | −0·88321 818 | +145 743 | 16·80 | +0·11709 695 | − 19 675 | +1·03825 247 | −173 286 |
| ·31 | ·18388 257 | 29 994 | ·85394 028 | 141 139 | ·81 | ·10661 897 | 17 928 | ·05705 274 | 176 652 |
| ·32 | ·19226 969 | 31 385 | ·82325 142 | 136 302 | ·82 | ·09596 178 | 16 141 | ·07408 702 | 179 733 |
| ·33 | ·20034 306 | 32 720 | ·79119 997 | 131 228 | ·83 | ·08514 322 | 14 335 | ·08932 452 | 182 511 |
| ·34 | ·20808 932 | 34 008 | ·75783 665 | 125 933 | ·84 | ·07418 137 | 12 492 | ·10273 746 | 184 991 |
| 16·35 | +0·21549 560 | − 35 239 | −0·72321 440 | +120 423 | 16·85 | +0·06309 463 | − 10 634 | +1·11430 106 | −187 159 |
| ·36 | ·22254 959 | 36 415 | ·68738 831 | 114 701 | ·86 | ·05190 159 | 8 752 | ·12399 364 | 189 020 |
| ·37 | ·22923 954 | 37 532 | ·65041 557 | 108 785 | ·87 | ·04062 106 | 6 855 | ·13179 661 | 190 560 |
| ·38 | ·23555 428 | 38 591 | ·61235 533 | 102 678 | ·88 | ·02927 201 | 4 942 | ·13769 457 | 191 782 |
| ·39 | ·24148 323 | 39 586 | ·57326 864 | 96 390 | ·89 | ·01787 356 | 3 021 | ·14167 530 | 192 683 |
| 16·40 | +0·24701 644 | − 40 517 | −0·53321 836 | + 89 933 | 16·90 | +0·00644 492 | − 1 088 | +1·14372 980 | −193 259 |
| ·41 | ·25214 460 | 41 385 | ·49226 904 | 83 319 | ·91 | − ·00499 460 | + 845 | ·14385 231 | 193 514 |
| ·42 | ·25685 904 | 42 182 | ·45048 681 | 76 552 | ·92 | ·01642 567 | 2 778 | ·14204 034 | 193 430 |
| ·43 | ·26115 178 | 42 914 | ·40793 931 | 69 648 | ·93 | ·02782 896 | 4 712 | ·13829 467 | 193 026 |
| ·44 | ·26501 551 | 43 576 | ·36469 556 | 62 617 | ·94 | ·03918 514 | 6 640 | ·13261 935 | 192 288 |
| 16·45 | +0·26844 361 | − 44 166 | −0·32082 585 | + 55 473 | 16·95 | −0·05047 494 | + 8 556 | +1·12502 174 | −191 231 |
| ·46 | ·27143 018 | 44 686 | ·27640 161 | 48 217 | ·96 | ·06167 920 | 10 463 | ·11551 243 | 189 839 |
| ·47 | ·27397 003 | 45 129 | ·23149 536 | 40 876 | ·97 | ·07277 886 | 12 352 | ·10410 532 | 188 126 |
| ·48 | ·27605 872 | 45 502 | ·18618 050 | 33 449 | ·98 | ·08375 503 | 14 223 | ·09081 754 | 186 091 |
| ·49 | ·27769 253 | 45 800 | ·14053 127 | 25 956 | ·99 | ·09458 901 | 16 074 | ·07566 944 | 183 734 |
| 16·50 | +0·27886 848 | − 46 021 | −0·09462 258 | + 18 406 | 17·00 | −0·10526 230 | + 17 897 | +1·05868 458 | −181 061 |

TABLE I—Ai(x) AND Ai'(x)

| x | Ai(−x) | $\delta_m^2$ | Ai'(−x) | $\delta_m^2$ | x | Ai(−x) | $\delta_m^2$ | Ai'(−x) | $\delta_m^2$ |
|---|---|---|---|---|---|---|---|---|---|
| 17·00 | −0·10526 230 | +17 897 | +1·05868 458 | −181 061 | 17·50 | −0·17266 059 | +30 221 | −0·90240 492 | +156 220 |
| ·01 | ·11575 667 | 19 694 | ·03988 969 | 178 074 | ·51 | ·16348 809 | 28 633 | ·93183 055 | 161 557 |
| ·02 | ·12605 416 | 21 457 | 1·01931 463 | 174 779 | ·52 | ·15402 936 | 26 989 | ·95964 112 | 166 617 |
| ·03 | ·13613 714 | 23 190 | 0·99699 234 | 171 177 | ·53 | ·14430 082 | 25 302 | 0·98578 604 | 171 395 |
| ·04 | ·14598 830 | 24 879 | ·97295 882 | 167 284 | ·54 | ·13431 935 | 23 562 | 1·01021 755 | 175 880 |
| 17·05 | −0·15559 074 | +26 533 | +0·94725 301 | −163 090 | 17·55 | −0·12410 233 | +21 785 | −1·03289 082 | +180 062 |
| ·06 | ·16492 793 | 28 141 | ·91991 682 | 158 616 | ·56 | ·11366 754 | 19 964 | ·05376 404 | 183 936 |
| ·07 | ·17398 379 | 29 704 | ·89099 499 | 153 860 | ·57 | ·10303 318 | 18 106 | ·07279 848 | 187 494 |
| ·08 | ·18274 270 | 31 217 | ·86053 506 | 148 833 | ·58 | ·09221 782 | 16 216 | ·08995 858 | 190 724 |
| ·09 | ·19118 953 | 32 680 | ·82858 728 | 143 544 | ·59 | ·08124 036 | 14 291 | ·10521 204 | 193 629 |
| 17·10 | −0·19930 966 | +34 087 | +0·79520 453 | −137 995 | 17·60 | −0·07012 003 | +12 345 | −1·11852 983 | +196 195 |
| ·11 | ·20708 902 | 35 439 | ·76044 227 | 132 208 | ·61 | ·05887 630 | 10 369 | ·12988 630 | 198 418 |
| ·12 | ·21451 410 | 36 732 | ·72435 837 | 126 177 | ·62 | ·04752 891 | 8 376 | ·13925 922 | 200 298 |
| ·13 | ·22157 198 | 37 960 | ·68701 311 | 119 923 | ·63 | ·03609 779 | 6 366 | ·14662 981 | 201 825 |
| ·14 | ·22825 037 | 39 131 | ·64846 902 | 113 451 | ·64 | ·02460 304 | 4 341 | ·15198 280 | 203 000 |
| 17·15 | −0·23453 758 | +40 227 | +0·60879 080 | −106 772 | 17·65 | −0·01306 490 | + 2 306 | −1·15530 645 | +203 817 |
| ·16 | ·24042 263 | 41 266 | ·56804 522 | 99 900 | ·66 | − ·00150 371 | + 267 | ·15659 259 | 204 277 |
| ·17 | ·24589 516 | 42 227 | ·52630 098 | 92 843 | ·67 | + ·01006 014 | − 1 779 | ·15583 663 | 204 372 |
| ·18 | ·25094 555 | 43 119 | ·48362 863 | 85 612 | ·68 | ·02160 621 | 3 820 | ·15303 761 | 204 110 |
| ·19 | ·25556 488 | 43 940 | ·44010 045 | 78 224 | ·69 | ·03311 409 | 5 858 | ·14819 816 | 203 485 |
| 17·20 | −0·25974 495 | +44 682 | +0·39579 030 | − 70 687 | 17·70 | +0·04456 340 | − 7 891 | −1·14132 453 | +202 495 |
| ·21 | ·26347 833 | 45 354 | ·35077 353 | 63 014 | ·71 | ·05593 383 | 9 905 | ·13242 661 | 201 150 |
| ·22 | ·26675 832 | 45 942 | ·30512 684 | 55 222 | ·72 | ·06720 523 | 11 912 | ·12151 786 | 199 441 |
| ·23 | ·26957 903 | 46 458 | ·25892 813 | 47 318 | ·73 | ·07835 755 | 13 895 | ·10861 536 | 197 376 |
| ·24 | ·27193 531 | 46 890 | ·21225 641 | 39 321 | ·74 | ·08937 096 | 15 858 | ·09373 975 | 194 960 |
| 17·25 | −0·27382 284 | +47 242 | +0·16519 163 | − 31 241 | 17·75 | +0·10022 584 | −17 792 | −1·07691 519 | +192 188 |
| ·26 | ·27523 810 | 47 514 | ·11781 457 | 23 092 | ·76 | ·11090 285 | 19 699 | ·05816 938 | 189 075 |
| ·27 | ·27617 837 | 47 705 | ·07020 669 | 14 892 | ·77 | ·12138 293 | 21 574 | ·03753 345 | 185 616 |
| ·28 | ·27664 174 | 47 812 | + ·02244 997 | − 6 647 | ·78 | ·13164 734 | 23 411 | 1·01504 197 | 181 825 |
| ·29 | ·27662 714 | 47 836 | − ·02537 318 | + 1 620 | ·79 | ·14167 771 | 25 209 | 0·99073 385 | 177 700 |
| 17·30 | −0·27613 433 | +47 781 | −0·07318 011 | + 9 898 | 17·80 | +0·15145 607 | −26 963 | −0·96464 732 | +173 253 |
| ·31 | ·27516 387 | 47 639 | ·12088 806 | 18 176 | ·81 | ·16096 488 | 28 672 | ·93682 984 | 168 489 |
| ·32 | ·27371 717 | 47 415 | ·16841 428 | 26 435 | ·82 | ·17018 706 | 30 334 | ·90732 803 | 163 420 |
| ·33 | ·27179 647 | 47 111 | ·21567 621 | 34 663 | ·83 | ·17910 600 | 31 940 | ·87619 258 | 158 044 |
| ·34 | ·26940 481 | 46 724 | ·26259 159 | 42 846 | ·84 | ·18770 564 | 33 493 | ·84347 722 | 152 382 |
| 17·35 | −0·26654 606 | +46 252 | −0·30907 862 | + 50 966 | 17·85 | +0·19597 046 | −34 986 | −0·80923 856 | +146 436 |
| ·36 | ·26322 493 | 45 704 | ·35505 612 | 59 014 | ·86 | ·20388 553 | 36 421 | ·77353 604 | 140 219 |
| ·37 | ·25944 691 | 45 075 | ·40044 364 | 66 973 | ·87 | ·21143 651 | 37 790 | ·73643 181 | 133 740 |
| ·38 | ·25521 829 | 44 364 | ·44516 162 | 74 828 | ·88 | ·21860 971 | 39 095 | ·69799 064 | 127 009 |
| ·39 | ·25054 617 | 43 578 | ·48913 153 | 82 568 | ·89 | ·22539 209 | 40 329 | ·65827 981 | 120 044 |
| 17·40 | −0·24543 841 | +42 714 | −0·53227 600 | + 90 177 | 17·90 | +0·23177 131 | −41 496 | −0·61736 896 | +112 848 |
| ·41 | ·23990 365 | 41 773 | ·57451 897 | 97 640 | ·91 | ·23773 571 | 42 586 | ·57533 002 | 105 439 |
| ·42 | ·23395 129 | 40 764 | ·61578 583 | 104 947 | ·92 | ·24327 439 | 43 602 | ·53223 706 | 97 829 |
| ·43 | ·22759 143 | 39 674 | ·65600 353 | 112 085 | ·93 | ·24837 719 | 44 541 | ·48816 616 | 90 029 |
| ·44 | ·22083 495 | 38 521 | ·69510 072 | 119 036 | ·94 | ·25303 472 | 45 405 | ·44319 529 | 82 057 |
| 17·45 | −0·21369 339 | +37 298 | −0·73300 791 | +125 795 | 17·95 | +0·25723 836 | −46 180 | −0·39740 415 | + 73 921 |
| ·46 | ·20617 898 | 36 003 | ·76965 754 | 132 342 | ·96 | ·26098 034 | 46 882 | ·35087 407 | 65 639 |
| ·47 | ·19830 465 | 34 652 | ·80498 415 | 138 672 | ·97 | ·26425 366 | 47 494 | ·30368 784 | 57 228 |
| ·48 | ·19008 392 | 33 231 | ·83892 447 | 144 766 | ·98 | ·26705 219 | 48 025 | ·25592 955 | 48 697 |
| ·49 | ·18153 098 | 31 756 | ·87141 757 | 150 622 | ·99 | ·26937 063 | 48 469 | ·20768 448 | 40 065 |
| 17·50 | −0·17266 059 | +30 221 | −0·90240 492 | +156 220 | 18·00 | +0·27120 454 | −48 825 | −0·15903 892 | + 31 348 |

## TABLE I—Ai($x$) AND Ai$'$($x$)

| $x$ | Ai($-x$) | $\delta^2_m$ | Ai$'$($-x$) | $\delta^2_m$ | $x$ | Ai($-x$) | $\delta^2_m$ | Ai$'$($-x$) | $\delta^2_m$ |
|---|---|---|---|---|---|---|---|---|---|
| 18·00 | +0·27120 454 | −48 825 | −0·15903 892 | + 31 348 | 18·50 | −0·11208 854 | +20 741 | +1·06464 396 | −198 116 |
| ·01 | ·27255 036 | 49 095 | ·11008 002 | 22 554 | ·51 | ·12262 801 | 22 702 | ·04292 338 | 194 310 |
| ·02 | ·27340 539 | 49 276 | ·06089 567 | 13 713 | ·52 | ·13294 053 | 24 625 | 1·01926 038 | 190 132 |
| ·03 | ·27376 782 | 49 371 | − ·01157 427 | + 4 826 | ·53 | ·14300 688 | 26 504 | 0·99369 672 | 185 598 |
| ·04 | ·27363 671 | 49 370 | + ·03779 535 | − 4 080 | ·54 | ·15280 828 | 28 335 | ·96627 773 | 180 710 |
| 18·05 | +0·27301 205 | −49 291 | +0·08712 415 | − 12 996 | 18·55 | −0·16232 642 | +30 118 | +0·93705 227 | −175 479 |
| ·06 | ·27189 466 | 49 110 | ·13632 300 | 21 905 | ·56 | ·17154 348 | 31 843 | ·90607 263 | 169 916 |
| ·07 | ·27028 632 | 48 852 | ·18530 285 | 30 784 | ·57 | ·18044 221 | 33 516 | ·87339 443 | 164 026 |
| ·08 | ·26818 963 | 48 495 | ·23397 493 | 39 627 | ·58 | ·18900 590 | 35 122 | ·83907 655 | 157 821 |
| ·09 | ·26560 814 | 48 059 | ·28225 085 | 48 410 | ·59 | ·19721 848 | 36 671 | ·80318 102 | 151 313 |
| 18·10 | +0·26254 623 | −47 528 | +0·33004 280 | − 57 121 | 18·60 | −0·20506 448 | +38 149 | +0·76577 290 | −144 513 |
| ·11 | ·25900 919 | 46 916 | ·37726 370 | 65 744 | ·61 | ·21252 912 | 39 560 | ·72692 017 | 137 432 |
| ·12 | ·25500 315 | 46 216 | ·42382 736 | 74 259 | ·62 | ·21959 830 | 40 895 | ·68669 361 | 130 084 |
| ·13 | ·25053 511 | 45 429 | ·46964 865 | 82 656 | ·63 | ·22625 866 | 42 161 | ·64516 668 | 122 481 |
| ·14 | ·24561 293 | 44 563 | ·51464 363 | 90 914 | ·64 | ·23249 756 | 43 346 | ·60241 538 | 114 639 |
| 18·15 | +0·24024 527 | −43 613 | +0·55872 974 | − 99 025 | 18·65 | −0·23830 315 | +44 452 | +0·55851 811 | −106 566 |
| ·16 | ·23444 163 | 42 581 | ·60182 591 | 106 964 | ·66 | ·24366 437 | 45 476 | ·51355 556 | 98 286 |
| ·17 | ·22821 232 | 41 476 | ·64385 277 | 114 726 | ·67 | ·24857 098 | 46 417 | ·46761 051 | 89 808 |
| ·18 | ·22156 840 | 40 287 | ·68473 273 | 122 291 | ·68 | ·25301 358 | 47 271 | ·42076 772 | 81 145 |
| ·19 | ·21452 174 | 39 030 | ·72439 017 | 129 643 | ·69 | ·25698 363 | 48 040 | ·37311 378 | 72 321 |
| 18·20 | +0·20708 492 | −37 697 | +0·76275 159 | − 136 774 | 18·70 | −0·26047 345 | +48 718 | +0·32473 691 | − 63 343 |
| ·21 | ·19927 126 | 36 293 | ·79974 571 | 143 666 | ·71 | ·26347 626 | 49 306 | ·27572 685 | 54 236 |
| ·22 | ·19109 479 | 34 824 | ·83530 363 | 150 309 | ·72 | ·26598 618 | 49 802 | ·22617 465 | 45 009 |
| ·23 | ·18257 020 | 33 290 | ·86935 895 | 156 685 | ·73 | ·26799 825 | 50 205 | ·17617 254 | 35 685 |
| ·24 | ·17371 283 | 31 689 | ·90184 792 | 162 789 | ·74 | ·26950 844 | 50 515 | ·12581 373 | 26 280 |
| 18·25 | +0·16453 867 | −30 036 | +0·93270 953 | − 168 604 | 18·75 | −0·27051 365 | +50 730 | +0·07519 224 | − 16 807 |
| ·26 | ·15506 426 | 28 318 | ·96188 565 | 174 121 | ·76 | ·27101 173 | 50 853 | + ·02440 276 | − 7 293 |
| ·27 | ·14530 676 | 26 554 | 0·98932 113 | 179 329 | ·77 | ·27100 146 | 50 875 | − ·02645 959 | + 2 258 |
| ·28 | ·13528 382 | 24 734 | 1·01496 391 | 184 214 | ·78 | ·27048 261 | 50 807 | ·07729 935 | 11 810 |
| ·29 | ·12501 362 | 22 870 | ·03876 515 | 188 776 | ·79 | ·26945 587 | 50 641 | ·12802 101 | 21 364 |
| 18·30 | +0·11451 480 | −20 959 | +1·06067 926 | − 192 994 | 18·80 | −0·26792 290 | +50 378 | −0·17852 908 | + 30 889 |
| ·31 | ·10380 646 | 19 011 | ·08066 407 | 196 866 | ·81 | ·26588 632 | 50 024 | ·22872 834 | 40 371 |
| ·32 | ·09290 808 | 17 024 | ·09868 087 | 200 387 | ·82 | ·26334 968 | 49 571 | ·27852 400 | 49 793 |
| ·33 | ·08183 952 | 15 003 | ·11469 447 | 203 542 | ·83 | ·26031 750 | 49 029 | ·32782 188 | 59 134 |
| ·34 | ·07062 098 | 12 957 | ·12867 333 | 206 330 | ·84 | ·25679 521 | 48 389 | ·37652 859 | 68 382 |
| 18·35 | +0·05927 293 | −10 877 | +1·14058 958 | − 208 745 | 18·85 | −0·25278 920 | +47 660 | −0·42455 169 | + 77 513 |
| ·36 | ·04781 614 | 8 782 | ·15041 908 | 210 776 | ·86 | ·24830 676 | 46 839 | ·47179 990 | 86 513 |
| ·37 | ·03627 157 | 6 663 | ·15814 152 | 212 426 | ·87 | ·24335 609 | 45 930 | ·51818 325 | 95 365 |
| ·38 | ·02466 039 | 4 534 | ·16374 041 | 213 689 | ·88 | ·23794 628 | 44 934 | ·56361 326 | 104 050 |
| ·39 | ·01300 389 | 2 392 | ·16720 313 | 214 558 | ·89 | ·23208 729 | 43 849 | ·60800 311 | 112 549 |
| 18·40 | +0·00132 348 | − 245 | +1·16852 099 | − 215 033 | 18·90 | −0·22578 996 | +42 682 | −0·65126 783 | +120 855 |
| ·41 | − ·01035 937 | + 1 909 | ·16768 924 | 215 115 | ·91 | ·21906 596 | 41 434 | ·69332 440 | 128 939 |
| ·42 | ·02202 314 | 4 057 | ·16470 707 | 214 800 | ·92 | ·21192 777 | 40 104 | ·73409 200 | 136 795 |
| ·43 | ·03364 635 | 6 201 | ·15957 763 | 214 085 | ·93 | ·20438 868 | 38 699 | ·77349 210 | 144 406 |
| ·44 | ·04520 756 | 8 339 | ·15230 806 | 212 978 | ·94 | ·19646 274 | 37 216 | ·81144 863 | 151 750 |
| 18·45 | −0·05668 541 | + 10 459 | +1·14290 944 | − 211 473 | 18·95 | −0·18816 477 | +35 667 | −0·84788 816 | +158 824 |
| ·46 | ·06805 870 | 12 566 | ·13139 681 | 209 576 | ·96 | ·17951 027 | 34 041 | ·88273 999 | 165 601 |
| ·47 | ·07930 637 | 14 650 | ·11778 914 | 207 287 | ·97 | ·17051 548 | 32 352 | ·91593 636 | 172 082 |
| ·48 | ·09040 758 | 16 712 | ·10210 931 | 204 612 | ·98 | ·16119 728 | 30 603 | ·94741 250 | 178 237 |
| ·49 | ·10134 173 | 18 739 | ·08438 406 | 201 555 | ·99 | ·15157 317 | 28 788 | 0·97710 686 | 184 074 |
| 18·50 | −0·11208 854 | + 20 741 | +1·06464 396 | − 198 116 | 19·00 | −0·14166 128 | +26 922 | −1·00496 112 | +189 559 |

## TABLE I—Ai($x$) and Ai'($x$)

| $x$ | Ai($-x$) | $\delta_m^2$ | Ai'($-x$) | $\delta_m^2$ | $x$ | Ai($-x$) | $\delta_m^2$ | Ai'($-x$) | $\delta_m^2$ |
|---|---|---|---|---|---|---|---|---|---|
| 19·00 | −0·14166 128 | +26 922 | −1·00496 112 | +189 559 | 19·50 | +0·26780 027 | − 52 230 | +0·08774 109 | −  14 434 |
| ·01 | ·13148 027 | 24 999 | ·03092 043 | 194 701 | ·51 | ·26666 204 | 52 037 | ·13987 299 | 24 626 |
| ·02 | ·12104 936 | 23 028 | ·05493 340 | 199 475 | ·52 | ·26500 363 | 51 739 | ·19175 869 | 34 785 |
| ·03 | ·11038 825 | 21 012 | ·07695 230 | 203 878 | ·53 | ·26282 802 | 51 339 | ·24329 663 | 44 897 |
| ·04 | ·09951 710 | 18 951 | ·09693 312 | 207 900 | ·54 | ·26013 920 | 50 843 | ·29438 574 | 54 929 |
| 19·05 | −0·08845 651 | +16 854 | −1·11483 566 | +211 533 | 19·55 | +0·25694 214 | −50 242 | +0·34492 572 | −  64 875 |
| ·06 | ·07722 744 | 14 723 | ·13062 361 | 214 764 | ·56 | ·25324 284 | 49 542 | ·39481 716 | 74 707 |
| ·07 | ·06585 120 | 12 561 | ·14426 466 | 217 595 | ·57 | ·24904 829 | 48 751 | ·44396 177 | 84 409 |
| ·08 | ·05434 940 | 10 372 | ·15573 052 | 220 013 | ·58 | ·24436 642 | 47 855 | ·49226 257 | 93 958 |
| ·09 | ·04274 392 | 8 161 | ·16499 702 | 222 011 | ·59 | ·23920 617 | 46 872 | ·53962 410 | 103 338 |
| 19·10 | −0·03105 686 | + 5 934 | −1·17204 418 | +223 595 | 19·60 | +0·23357 738 | −45 789 | +0·58595 259 | −112 532 |
| ·11 | ·01931 049 | 3 692 | ·17685 618 | 224 745 | ·61 | ·22749 086 | 44 619 | ·63115 614 | 121 517 |
| ·12 | − ·00752 722 | + 1 439 | ·17942 151 | 225 475 | ·62 | ·22095 831 | 43 361 | ·67514 493 | 130 279 |
| ·13 | + ·00427 043 | − 817 | ·17973 289 | 225 768 | ·63 | ·21399 231 | 42 015 | ·71783 138 | 138 795 |
| ·14 | ·01605 991 | 3 073 | ·17778 738 | 225 634 | ·64 | ·20660 631 | 40 586 | ·75913 035 | 147 057 |
| 19·15 | +0·02781 866 | − 5 329 | −1·17358 633 | +225 062 | 19·65 | +0·19881 460 | −39 074 | +0·79895 927 | −155 035 |
| ·16 | ·03952 414 | 7 574 | ·16713 545 | 224 063 | ·66 | ·19063 229 | 37 486 | ·83723 837 | 162 727 |
| ·17 | ·05115 390 | 9 807 | ·15844 474 | 222 628 | ·67 | ·18207 526 | 35 823 | ·87389 077 | 170 105 |
| ·18 | ·06268 562 | 12 026 | ·14752 854 | 220 767 | ·68 | ·17316 014 | 34 083 | ·90884 271 | 177 164 |
| ·19 | ·07409 712 | 14 223 | ·13440 546 | 218 476 | ·69 | ·16390 431 | 32 281 | ·94202 364 | 183 879 |
| 19·20 | +0·08536 644 | −16 392 | −1·11909 840 | +215 762 | 19·70 | +0·15432 580 | −30 408 | +0·97336 642 | −190 248 |
| ·21 | ·09647 189 | 18 535 | ·10163 449 | 212 632 | ·71 | ·14444 332 | 28 476 | 1·00280 740 | 196 248 |
| ·22 | ·10739 205 | 20 645 | ·08204 503 | 209 082 | ·72 | ·13427 619 | 26 485 | ·03028 660 | 201 869 |
| ·23 | ·11810 583 | 22 717 | ·06036 549 | 205 132 | ·73 | ·12384 431 | 24 439 | ·05574 783 | 207 099 |
| ·24 | ·12859 252 | 24 745 | ·03663 538 | 200 773 | ·74 | ·11316 813 | 22 343 | ·07913 880 | 211 933 |
| 19·25 | +0·13883 184 | −26 728 | −1·01089 826 | +196 027 | 19·75 | +0·10226 860 | −20 203 | +1·10041 120 | −216 352 |
| ·26 | ·14880 396 | 28 668 | 0·98320 159 | 190 888 | ·76 | ·09116 712 | 18 019 | ·11952 086 | 220 348 |
| ·27 | ·15848 951 | 30 545 | ·95359 672 | 185 381 | ·77 | ·07988 552 | 15 796 | ·13642 783 | 223 918 |
| ·28 | ·16786 971 | 32 371 | ·92213 872 | 179 504 | ·78 | ·06844 602 | 13 541 | ·15109 643 | 227 045 |
| ·29 | ·17692 631 | 34 138 | ·88888 634 | 173 269 | ·79 | ·05687 116 | 11 258 | ·16349 539 | 229 735 |
| 19·30 | +0·18564 166 | −35 834 | −0·85390 190 | +166 693 | 19·80 | +0·04518 377 | − 8 949 | +1·17359 784 | −231 966 |
| ·31 | ·19399 879 | 37 470 | ·81725 114 | 159 784 | ·81 | ·03340 693 | 6 619 | ·18138 147 | 233 744 |
| ·32 | ·20198 136 | 39 029 | ·77900 313 | 152 553 | ·82 | ·02156 393 | 4 275 | ·18682 851 | 235 060 |
| ·33 | ·20957 377 | 40 520 | ·73923 015 | 145 019 | ·83 | + ·00967 820 | − 1 920 | ·18992 580 | 235 913 |
| ·34 | ·21676 113 | 41 928 | ·69800 752 | 137 190 | ·84 | − ·00222 672 | + 440 | ·19066 482 | 236 296 |
| 19·35 | +0·22352 935 | −43 263 | −0·65541 350 | +129 084 | 19·85 | −0·01412 723 | + 2 806 | +1·18904 174 | −236 215 |
| ·36 | ·22986 510 | 44 509 | ·61152 912 | 120 715 | ·86 | ·02599 969 | 5 164 | ·18505 738 | 235 659 |
| ·37 | ·23575 591 | 45 675 | ·56643 804 | 112 101 | ·87 | ·03782 052 | 7 516 | ·17871 729 | 234 638 |
| ·38 | ·24119 013 | 46 753 | ·52022 638 | 103 252 | ·88 | ·04956 621 | 9 855 | ·17003 169 | 233 143 |
| ·39 | ·24615 699 | 47 737 | ·47298 259 | 94 194 | ·89 | ·06121 338 | 12 177 | ·15901 551 | 231 188 |
| 19·40 | +0·25064 664 | −48 635 | −0·42479 723 | + 84 935 | 19·90 | −0·07273 882 | +14 478 | +1·14568 831 | −228 765 |
| ·41 | ·25465 011 | 49 438 | ·37576 285 | 75 497 | ·91 | ·08411 953 | 16 752 | ·13007 430 | 225 886 |
| ·42 | ·25815 938 | 50 143 | ·32597 379 | 65 902 | ·92 | ·09533 278 | 18 993 | ·11220 227 | 222 547 |
| ·43 | ·26116 739 | 50 755 | ·27552 598 | 56 157 | ·93 | ·10635 616 | 21 202 | ·09210 559 | 218 767 |
| ·44 | ·26366 803 | 51 267 | ·22451 682 | 46 295 | ·94 | ·11716 760 | 23 367 | ·06982 206 | 214 536 |
| 19·45 | +0·26565 618 | −51 679 | −0·17304 491 | + 36 321 | 19·95 | −0·12774 545 | +25 490 | +1·04539 396 | −209 877 |
| ·46 | ·26712 772 | 51 994 | ·12120 994 | 26 265 | ·96 | ·13806 849 | 27 565 | 1·01886 788 | 204 789 |
| ·47 | ·26807 951 | 52 206 | ·06911 244 | 16 142 | ·97 | ·14811 598 | 29 583 | 0·99029 468 | 199 285 |
| ·48 | ·26850 943 | 52 316 | − ·01685 361 | + 5 971 | ·98 | ·15786 774 | 31 548 | ·95972 938 | 193 371 |
| ·49 | ·26841 638 | 52 325 | + ·03546 488 | − 4 226 | ·99 | ·16730 413 | 33 451 | ·92723 109 | 187 064 |
| 19·50 | +0·26780 027 | −52 230 | +0·08774 109 | − 14 434 | 20·00 | −0·17640 613 | +35 289 | +0·89286 286 | −180 376 |

TABLE II—$\log_{10} \mathrm{Ai}(x)$ AND $\mathrm{Ai}'(x)/\mathrm{Ai}(x)$

| $x$ | $\log_{10} \mathrm{Ai}(x)$ | $\delta^2_m$ | $\mathrm{Ai}'(x)/\mathrm{Ai}(x)$ | $\delta^2_m$ | $x$ | $\log_{10} \mathrm{Ai}(x)$ | $\delta^2_m$ | $\mathrm{Ai}'(x)/\mathrm{Ai}(x)$ | $\delta^2$ |
|---|---|---|---|---|---|---|---|---|---|
| 0·0 | $\overline{1}$·55026 267 | − 230 703 | − 0·72901 11 | + 224 74 | 5·0 | $\overline{4}$·03480 658 | − 93 270 | − 2·28358 67 | + 19 13 |
| 0·1 | ·51746 396 | 221 433 | ·78106 92 | 202 90 | 5·1 | $\overline{5}$·93516 669 | 92 453 | ·30496 87 | 18 63 |
| 0·2 | ·48244 930 | 213 036 | ·83109 27 | 184 08 | 5·2 | ·83460 224 | 91 651 | ·32616 44 | 18 17 |
| 0·3 | ·44530 287 | 205 406 | ·87927 07 | 167 79 | 5·3 | ·73312 124 | 90 871 | ·34717 84 | 17 73 |
| 0·4 | ·40610 116 | 198 433 | ·92576 69 | 153 60 | 5·4 | ·63073 149 | 90 114 | ·36801 51 | 17 30 |
| 0·5 | ·36491 405 | − 192 039 | − 0·97072 39 | + 141 12 | 5·5 | ·52744 057 | − 89 370 | − 2·38867 88 | + 16 88 |
| 0·6 | ·32180 561 | 186 152 | 1·01426 69 | 130 17 | 5·6 | ·42325 592 | 88 645 | ·40917 37 | 16 49 |
| 0·7 | ·27683 482 | 180 716 | ·05650 59 | 120 45 | 5·7 | ·31818 479 | 87 940 | ·42950 37 | 16 11 |
| 0·8 | ·23005 614 | 175 677 | ·09753 84 | 111 79 | 5·8 | $\overline{2}$1223 424 | 87 244 | ·44967 26 | 15 74 |
| 0·9 | ·18152 004 | 170 990 | ·13745 13 | 104 07 | 5·9 | $\overline{5}$·10541 121 | 86 573 | ·46968 41 | 15 39 |
| 1·0 | ·13127 345 | − 166 624 | − 1·17632 20 | + 97 16 | 6·0 | $\overline{6}$·99772 243 | − 85 907 | − 2·48954 17 | + 15 04 |
| 1·1 | ·07936 009 | 162 544 | ·21421 99 | 90 88 | 6·1 | ·88917 454 | 85 265 | ·50924 89 | 14 73 |
| 1·2 | $\overline{1}$·02582 082 | 158 720 | ·25120 78 | 85 28 | 6·2 | ·77977 398 | 84 631 | ·52880 88 | 14 40 |
| 1·3 | $\overline{2}$·97069 392 | 155 130 | ·28734 20 | 80 14 | 6·3 | ·66952 708 | 84 013 | ·54822 47 | 14 10 |
| 1·4 | ·91401 533 | 151 753 | ·32267 39 | 75 49 | 6·4 | ·55844 003 | 83 407 | ·56749 96 | 13 79 |
| 1·5 | ·85581 886 | − 148 567 | − 1·35725 01 | + 71 27 | 6·5 | ·44651 889 | − 82 814 | − 2·58663 66 | + 13 53 |
| 1·6 | ·79613 640 | 145 556 | ·39111 30 | 67 37 | 6·6 | ·33376 959 | 82 233 | ·60563 83 | 13 23 |
| 1·7 | ·73499 808 | 142 712 | ·42430 16 | 63 83 | 6·7 | ·22019 794 | 81 664 | ·62450 77 | 12 98 |
| 1·8 | ·67243 238 | 140 008 | ·45685 14 | 60 54 | 6·8 | $\overline{6}$·10580 963 | 81 106 | ·64324 73 | 12 72 |
| 1·9 | ·60846 634 | 137 447 | ·48879 53 | 57 54 | 6·9 | $\overline{7}$·99061 024 | 80 559 | ·66185 97 | 12 46 |
| 2·0 | ·54312 560 | − 135 008 | − 1·52016 34 | + 54 75 | 7·0 | ·87460 524 | − 80 025 | − 2·68034 75 | + 12 22 |
| 2·1 | ·47643 456 | 132 688 | ·55098 36 | 52 18 | 7·1 | ·75779 998 | 79 495 | ·69871 31 | 12 00 |
| 2·2 | ·40841 644 | 130 475 | ·58128 17 | 49 79 | 7·2 | ·64019 974 | 78 984 | ·71695 87 | 11 76 |
| 2·3 | ·33909 339 | 128 361 | ·61108 16 | 47 56 | 7·3 | ·52180 965 | 78 475 | ·73508 67 | 11 54 |
| 2·4 | ·26848 656 | 126 341 | ·64040 56 | 45 49 | 7·4 | ·40263 479 | 77 980 | ·75309 93 | 11 33 |
| 2·5 | ·19661 616 | − 124 409 | − 1·66927 44 | + 43 58 | 7·5 | ·28268 012 | − 77 491 | − 2·77099 86 | + 11 12 |
| 2·6 | ·12350 153 | 122 552 | ·69770 72 | 41 75 | 7·6 | ·16195 052 | 77 014 | ·78878 67 | 10 92 |
| 2·7 | $\overline{2}$·04916 123 | 120 778 | ·72572 22 | 40 11 | 7·7 | $\overline{7}$·04045 077 | 76 544 | ·80646 56 | 10 73 |
| 2·8 | $\overline{3}$·97361 302 | 119 072 | ·75333 60 | 38 48 | 7·8 | $\overline{8}$·91818 557 | 76 082 | ·82403 72 | 10 54 |
| 2·9 | ·89687 397 | 117 432 | ·78056 47 | 37 03 | 7·9 | ·79515 954 | 75 627 | ·84150 34 | 10 36 |
| 3·0 | ·81896 049 | − 115 853 | − 1·80742 30 | + 35 64 | 8·0 | ·67137 722 | − 75 183 | − 2·85886 60 | + 10 16 |
| 3·1 | ·73988 837 | 114 338 | ·83392 48 | 34 33 | 8·1 | ·54684 306 | 74 742 | ·87612 70 | 10 01 |
| 3·2 | ·65967 278 | 112 871 | ·86008 32 | 33 07 | 8·2 | ·42156 146 | 74 315 | ·89328 79 | 9 84 |
| 3·3 | ·57832 838 | 111 461 | ·88591 07 | 31 93 | 8·3 | ·29553 670 | 73 889 | ·91035 04 | 9 66 |
| 3·4 | ·49586 928 | 110 098 | ·91141 88 | 30 80 | 8·4 | ·16877 303 | 73 474 | ·92731 63 | 9 51 |
| 3·5 | ·41230 911 | − 108 783 | − 1·93661 87 | + 29 80 | 8·5 | $\overline{8}$·04127 461 | − 73 064 | − 2·94418 71 | + 9 35 |
| 3·6 | ·32766 103 | 107 510 | ·96152 06 | 28 79 | 8·6 | $\overline{9}$·91304 554 | 72 662 | ·96096 44 | 9 21 |
| 3·7 | ·24193 777 | 106 282 | 1·98613 45 | 27 86 | 8·7 | ·78408 984 | 72 263 | ·97764 96 | 9 05 |
| 3·8 | ·15515 162 | 105 090 | 2·01046 97 | 26 99 | 8·8 | ·65441 149 | 71 875 | 2·99424 43 | 8 91 |
| 3·9 | $\overline{3}$·06731 450 | 103 937 | ·03453 50 | 26 12 | 8·9 | ·52401 438 | 71 491 | 3·01074 99 | 8 77 |
| 4·0 | $\overline{4}$·97843 794 | − 102 821 | − 2·05833 90 | + 25 35 | 9·0 | ·39290 235 | − 71 113 | − 3·02716 78 | + 8 64 |
| 4·1 | ·88853 311 | 101 734 | ·08188 95 | 24 57 | 9·1 | ·26107 918 | 70 741 | ·04349 93 | 8 51 |
| 4·2 | ·79761 087 | 100 686 | ·10519 42 | 23 84 | 9·2 | $\overline{9}$·12854 859 | 70 374 | ·05974 57 | 8 36 |
| 4·3 | ·70568 172 | 99 663 | ·12826 04 | 23 17 | 9·3 | $\overline{10}$·99531 425 | 70 013 | ·07590 85 | 8 25 |
| 4·4 | ·61275 588 | 98 674 | ·15109 49 | 22 49 | 9·4 | ·86137 977 | 69 658 | ·09198 88 | 8 13 |
| 4·5 | ·51884 325 | − 97 709 | − 2·17370 44 | + 21 88 | 9·5 | ·72674 870 | − 69 308 | − 3·10798 78 | + 8 00 |
| 4·6 | ·42395 348 | 96 772 | ·19609 51 | 21 28 | 9·6 | ·59142 454 | 68 963 | ·12390 68 | 7 89 |
| 4·7 | ·32809 594 | 95 863 | ·21827 30 | 20 68 | 9·7 | ·45541 074 | 68 622 | ·13974 69 | 7 77 |
| 4·8 | ·23127 973 | 94 975 | ·24024 40 | 20 16 | 9·8 | ·31871 071 | 68 288 | ·15550 93 | 7 66 |
| 4·9 | ·13351 373 | 94 111 | ·26201 34 | 19 60 | 9·9 | ·18132 779 | 67 957 | ·17119 51 | 7 55 |
| 5·0 | $\overline{4}$·03480 658 | − 93 270 | − 2·28358 67 | + 19 13 | 10·0 | $\overline{10}$·04326 529 | − 67 632 | − 3·18680 54 | + 7 44 |

# TABLE II—$\log_{10}\mathrm{Ai}(x)$ AND $\mathrm{Ai}'(x)/\mathrm{Ai}(x)$

| $x$ | $\log_{10}\mathrm{Ai}(x)$ | $\delta^2$ | $\mathrm{Ai}'(x)/\mathrm{Ai}(x)$ | $\delta^2$ | $x$ | $\log_{10}\mathrm{Ai}(x)$ | $\delta^2$ | $\mathrm{Ai}'(x)/\mathrm{Ai}(x)$ | $\delta^2$ |
|---|---|---|---|---|---|---|---|---|---|
| 10·0 | $\overline{10}$·04326 529 | −67 633 | −3·18680 54 | +7 44 | 15·0 | $\overline{18}$·33545 038 | −55 596 | −3·88947 52 | +4 17 |
| 10·1 | $\overline{11}$·90452 646 | 67 311 | ·20234 13 | 7 35 | 15·1 | $\overline{18}$·16625 494 | 55 418 | ·90225 61 | 4 11 |
| 10·2 | ·76511 452 | 66 995 | ·21780 37 | 7 24 | 15·2 | $\overline{19}$·99650 532 | 55 239 | ·91499 59 | 4 09 |
| 10·3 | ·62503 263 | 66 682 | ·23319 37 | 7 13 | 15·3 | ·82620 331 | 55 063 | ·92769 48 | 4 04 |
| 10·4 | ·48428 392 | 66 376 | ·24851 24 | 7 05 | 15·4 | ·65535 067 | 54 888 | ·94035 33 | 4 01 |
| 10·5 | ·34287 145 | −66 070 | −3·26376 06 | +6 95 | 15·5 | ·48394 915 | −54 714 | −3·95297 17 | +3 97 |
| 10·6 | ·20079 828 | 65 771 | ·27893 93 | 6 85 | 15·6 | ·31200 049 | 54 544 | ·96555 04 | 3 92 |
| 10·7 | $\overline{11}$·05806 740 | 65 477 | ·29404 95 | 6 76 | 15·7 | $\overline{19}$·13950 639 | 54 374 | ·97808 99 | 3 91 |
| 10·8 | $\overline{12}$·91468 175 | 65 182 | ·30909 21 | 6 68 | 15·8 | $\overline{20}$·96646 855 | 54 204 | 3·99059 03 | 3 85 |
| 10·9 | ·77064 428 | 64 897 | ·32406 79 | 6 59 | 15·9 | ·79288 867 | 54 038 | 4·00305 22 | 3 82 |
| 11·0 | ·62595 784 | −64 611 | −3·33897 78 | +6 50 | 16·0 | ·61876 841 | −53 873 | −4·01547 59 | +3 80 |
| 11·1 | ·48062 529 | 64 331 | ·35382 27 | 6 42 | 16·1 | ·44410 942 | 53 709 | ·02786 16 | 3 75 |
| 11·2 | ·33464 943 | 64 053 | ·36860 34 | 6 34 | 16·2 | ·26891 334 | 53 547 | ·04020 98 | 3 72 |
| 11·3 | ·18803 304 | 63 781 | ·38332 07 | 6 26 | 16·3 | $\overline{20}$·09318 179 | 53 385 | ·05252 08 | 3 69 |
| 11·4 | $\overline{12}$·04077 884 | 63 511 | ·39797 54 | 6 17 | 16·4 | $\overline{21}$·91691 639 | 53 227 | ·06479 49 | 3 65 |
| 11·5 | $\overline{13}$·89288 953 | −63 242 | −3·41256 84 | +6 11 | 16·5 | ·74011 872 | −53 068 | −4·07703 25 | +3 63 |
| 11·6 | ·74436 780 | 62 981 | ·42710 03 | 6 03 | 16·6 | ·56279 037 | 52 912 | ·08923 38 | 3 59 |
| 11·7 | ·59521 626 | 62 720 | ·44157 19 | 5 95 | 16·7 | ·38493 290 | 52 756 | ·10139 92 | 3 56 |
| 11·8 | ·44543 752 | 62 463 | ·45598 40 | 5 88 | 16·8 | ·20654 787 | 52 602 | ·11352 90 | 3 52 |
| 11·9 | ·29503 415 | 62 209 | ·47033 73 | 5 82 | 16·9 | $\overline{21}$·02763 682 | 52 450 | ·12562 36 | 3 51 |
| 12·0 | $\overline{13}$·14400 869 | −61 958 | −3·48463 24 | +5 73 | 17·0 | $\overline{22}$·84820 127 | −52 298 | −4·13768 31 | +3 47 |
| 12·1 | $\overline{14}$·99236 365 | 61 710 | ·49887 02 | 5 69 | 17·1 | ·66824 274 | 52 149 | ·14970 79 | 3 43 |
| 12·2 | ·84010 151 | 61 465 | ·51305 11 | 5 60 | 17·2 | ·48776 272 | 52 000 | ·16169 84 | 3 41 |
| 12·3 | ·68722 472 | 61 223 | ·52717 60 | 5 54 | 17·3 | ·30676 270 | 51 852 | ·17365 48 | 3 39 |
| 12·4 | ·53373 570 | 60 984 | ·54124 55 | 5 48 | 17·4 | $\overline{22}$·12524 416 | 51 706 | ·18557 73 | 3 35 |
| 12·5 | ·37963 684 | −60 747 | −3·55526 02 | +5 42 | 17·5 | $\overline{23}$·94320 856 | −51 561 | −4·19746 63 | +3 32 |
| 12·6 | ·22493 051 | 60 514 | ·56922 07 | 5 35 | 17·6 | ·76065 735 | 51 417 | ·20932 21 | 3 30 |
| 12·7 | $\overline{14}$·06961 904 | 60 282 | ·58312 77 | 5 29 | 17·7 | ·57759 197 | 51 275 | ·22114 49 | 3 28 |
| 12·8 | $\overline{15}$·91370 475 | 60 053 | ·59698 18 | 5 24 | 17·8 | ·39401 384 | 51 133 | ·23293 49 | 3 23 |
| 12·9 | ·75718 993 | 59 828 | ·61078 35 | 5 18 | 17·9 | ·30993 138 | 50 993 | ·24469 36 | 3 22 |
| 13·0 | ·60007 683 | −59 603 | −3·62453 34 | +5 11 | 18·0 | $\overline{23}$·02532 499 | −50 853 | −4·25641 81 | +3 20 |
| 13·1 | ·44236 770 | 59 384 | ·63823 22 | 5 06 | 18·1 | $\overline{24}$·84021 707 | 50 717 | ·26811 16 | 3 15 |
| 13·2 | ·28406 473 | 59 164 | ·65188 04 | 5 02 | 18·2 | ·65460 198 | 50 578 | ·27977 36 | 3 15 |
| 13·3 | $\overline{15}$·12517 012 | 58 947 | ·66547 84 | 4 94 | 18·3 | ·46848 111 | 50 443 | ·29140 41 | 3 11 |
| 13·4 | $\overline{16}$·96568 604 | 58 735 | ·67902 70 | 4 91 | 18·4 | ·28185 581 | 50 308 | ·30300 35 | 3 10 |
| 13·5 | ·80561 461 | −58 522 | −3·69252 65 | +4 84 | 18·5 | $\overline{24}$·09472 743 | −50 175 | −4·31457 19 | +3 05 |
| 13·6 | ·64495 796 | 58 313 | ·70597 76 | 4 80 | 18·6 | $\overline{25}$·90709 730 | 50 042 | ·32610 98 | 3 05 |
| 13·7 | ·48371 818 | 58 106 | ·71938 07 | 4 74 | 18·7 | ·71896 675 | 49 911 | ·33761 72 | 3 02 |
| 13·8 | ·32189 734 | 57 901 | ·73273 64 | 4 70 | 18·8 | ·53033 709 | 49 779 | ·34909 44 | 2 99 |
| 13·9 | $\overline{16}$·15949 749 | 57 697 | ·74604 51 | 4 65 | 18·9 | ·34120 964 | 49 651 | ·36054 17 | 2 97 |
| 14·0 | $\overline{17}$·99652 067 | −57 498 | −3·75930 73 | +4 59 | 19·0 | $\overline{25}$·15158 568 | −49 521 | −4·37195 93 | +2 95 |
| 14·1 | ·83296 887 | 57 297 | ·77252 36 | 4 56 | 19·1 | $\overline{26}$·96146 651 | 49 395 | ·38334 74 | 2 93 |
| 14·2 | ·66884 410 | 57 103 | ·78569 43 | 4 51 | 19·2 | ·77085 339 | 49 268 | ·39470 62 | 2 89 |
| 14·3 | ·50414 830 | 56 906 | ·79881 99 | 4 45 | 19·3 | ·57974 759 | 49 142 | ·40603 61 | 2 89 |
| 14·4 | ·33888 344 | 56 715 | ·81190 10 | 4 42 | 19·4 | ·38815 037 | 49 017 | ·41733 71 | 2 86 |
| 14·5 | ·17305 143 | −56 523 | −3·82493 79 | +4 38 | 19·5 | ·19606 298 | −48 894 | −4·42860 95 | +2 84 |
| 14·6 | $\overline{17}$·00665 419 | 56 335 | ·83793 10 | 4 32 | 19·6 | $\overline{26}$·00348 665 | 48 770 | ·43985 35 | 2 81 |
| 14·7 | $\overline{18}$·83969 360 | 56 147 | ·85088 09 | 4 28 | 19·7 | $\overline{27}$·81042 262 | 48 650 | ·45106 94 | 2 80 |
| 14·8 | ·67217 154 | 55 962 | ·86378 80 | 4 25 | 19·8 | ·61687 209 | 48 528 | ·46225 73 | 2 78 |
| 14·9 | ·50408 986 | 55 780 | ·87665 26 | 4 20 | 19·9 | ·42283 628 | 48 409 | ·47341 74 | 2 75 |
| 15·0 | $\overline{18}$·33545 038 | −55 596 | −3·88947 52 | +4 17 | 20·0 | $\overline{27}$·22831 638 | −48 288 | −4·48455 00 | +2 73 |

## TABLE II—$\log_{10}\mathrm{Ai}(x)$ AND $\mathrm{Ai}'(x)/\mathrm{Ai}(x)$

| $x$ | $\log_{10}\mathrm{Ai}(x)$ | $\delta^2$ | $\mathrm{Ai}'(x)/\mathrm{Ai}(x)$ | $\delta^2$ |
|---|---|---|---|---|
| 20·0 | $\overline{27}$·22831 638 | −48 288 | −4·48455 00 | +2 73 |
| 20·1 | $\overline{27}$·03331 360 | 48 170 | ·49565 53 | 2 72 |
| 20·2 | $\overline{28}$·83782 912 | 48 054 | ·50673 34 | 2 69 |
| 20·3 | $\overline{28}$·64186 410 | 47 936 | ·51778 46 | 2 68 |
| 20·4 | $\overline{28}$·44541 972 | 47 821 | ·52880 90 | 2 65 |
| 20·5 | $\overline{28}$·24849 713 | −47 705 | −4·53980 69 | +2 64 |
| 20·6 | $\overline{28}$·05109 749 | 47 592 | ·55077 84 | 2 62 |
| 20·7 | $\overline{29}$·85322 193 | 47 479 | ·56172 37 | 2 60 |
| 20·8 | $\overline{29}$·65487 158 | 47 365 | ·57264 30 | 2 58 |
| 20·9 | $\overline{29}$·45604 758 | 47 254 | ·58353 65 | 2 57 |
| 21·0 | $\overline{29}$·25675 104 | −47 144 | −4·59440 43 | +2 54 |
| 21·1 | $\overline{29}$·05698 306 | 47 033 | ·60524 67 | 2 52 |
| 21·2 | $\overline{30}$·85674 475 | 46 923 | ·61606 39 | 2 52 |
| 21·3 | $\overline{30}$·65603 721 | 46 815 | ·62685 59 | 2 49 |
| 21·4 | $\overline{30}$·45486 152 | 46 707 | ·63762 30 | 2 48 |
| 21·5 | $\overline{30}$·25321 876 | −46 601 | −4·64836 53 | +2 45 |
| 21·6 | $\overline{30}$·05110 999 | 46 493 | ·65908 31 | 2 45 |
| 21·7 | $\overline{31}$·84853 629 | 46 387 | ·66977 64 | 2 42 |
| 21·8 | $\overline{31}$·64549 872 | 46 284 | ·68044 55 | 2 41 |
| 21·9 | $\overline{31}$·44199 831 | 46 178 | ·69109 05 | 2 40 |
| 22·0 | $\overline{31}$·23803 612 | −46 075 | −4·70171 15 | +2 37 |
| 22·1 | $\overline{31}$·03361 318 | 45 972 | ·71230 88 | 2 36 |
| 22·2 | $\overline{32}$·82873 052 | 45 870 | ·72288 25 | 2 35 |
| 22·3 | $\overline{32}$·62338 916 | 45 769 | ·73343 27 | 2 32 |
| 22·4 | $\overline{32}$·41759 011 | 45 667 | ·74395 97 | 2 33 |
| 22·5 | $\overline{32}$·21133 439 | −45 567 | −4·75446 34 | +2 29 |
| 22·6 | $\overline{32}$·00462 300 | 45 468 | ·76494 42 | 2 29 |
| 22·7 | $\overline{33}$·79745 693 | 45 369 | ·77540 21 | 2·26 |
| 22·8 | $\overline{33}$·58983 717 | 45 271 | ·78583 74 | 2 27 |
| 22·9 | $\overline{33}$·38176 470 | 45 172 | ·79625 00 | 2 23 |
| 23·0 | $\overline{33}$·17324 051 | −45 077 | −4·80664 03 | +2 23 |
| 23·1 | $\overline{34}$·96426 555 | 44 979 | ·81700 83 | 2 21 |
| 23·2 | $\overline{34}$·75484 080 | 44 884 | ·82735 42 | 2 20 |
| 23·3 | $\overline{34}$·54496 721 | 44 789 | ·83767 81 | 2 18 |
| 23·4 | $\overline{34}$·33464 573 | 44 693 | ·84798 02 | 2 18 |
| 23·5 | $\overline{34}$·12387 732 | −44 601 | −4·85826 05 | +2 15 |
| 23·6 | $\overline{35}$·91266 290 | 44 507 | ·86851 93 | 2 14 |
| 23·7 | $\overline{35}$·70100 341 | 44 414 | ·87875 67 | 2 14 |
| 23·8 | $\overline{35}$·48889 978 | 44 321 | ·88897 27 | 2 11 |
| 23·9 | $\overline{35}$·27635 294 | 44 231 | ·89916 76 | 2 10 |
| 24·0 | $\overline{35}$·06336 379 | −44 138 | −4·90934 15 | +2 10 |
| 24·1 | $\overline{36}$·84993 326 | 44 049 | ·91949 44 | 2 07 |
| 24·2 | $\overline{36}$·63606 224 | 43 958 | ·92962 66 | 2 07 |
| 24·3 | $\overline{36}$·42175 164 | 43 870 | ·93973 81 | 2 06 |
| 24·4 | $\overline{36}$·20700 234 | 43 780 | ·94982 90 | 2 03 |
| 24·5 | $\overline{37}$·99181 524 | −43 691 | −4·95989 96 | +2 03 |
| 24·6 | $\overline{37}$·77619 123 | 43 605 | ·96994 99 | 2 02 |
| 24·7 | $\overline{37}$·56013 117 | 43 516 | ·97998 00 | 2 00 |
| 24·8 | $\overline{37}$·34363 595 | 43 430 | ·98999 01 | 2 00 |
| 24·9 | $\overline{37}$·12670 643 | 43 344 | 4·99998 02 | 1 97 |
| 25·0 | $\overline{38}$·90934 347 | −43 257 | −5·00995 06 | +1 97 |

| $x$ | $\log_{10}\mathrm{Ai}(x)$ | $\delta^2_m$ | $\gamma^4$ | $\mathrm{Ai}'(x)/\mathrm{Ai}(x)$ | $\delta^2_m$ |
|---|---|---|---|---|---|
| 25 | $\overline{38}$·90934 347 | −4325 279 | −5 | −5·00995 06 | +196 75 |
| 26 | $\overline{40}$·71206 108 | 4242 271 | 5 | ·10859 01 | 185 69 |
| 27 | $\overline{42}$·47234 754 | 4163 851 | 4 | ·20537 09 | 175 62 |
| 28 | $\overline{44}$·19098 781 | 4089 606 | 4 | ·30039 39 | 166 42 |
| 29 | $\overline{47}$·86872 498 | 4019 188 | 4 | ·39375 13 | 157 99 |
| 30 | $\overline{49}$·50626 382 | −3952 274 | −3 | −5·48552 75 | +150 27 |
| 31 | $\overline{51}$·10427 399 | 3888 583 | 3 | ·57579 99 | 143 13 |
| 32 | $\overline{54}$·66339 286 | 3827 866 | 3 | ·66464 00 | 136 56 |
| 33 | $\overline{56}$·18422 800 | 3769 905 | 3 | ·75211 36 | 130 46 |
| 34 | $\overline{59}$·66735 941 | 3714 485 | 2 | ·83828 18 | 124 81 |
| 35 | $\overline{61}$·11334 160 | −3661 441 | −2 | −5·92320 12 | +119 54 |
| 36 | $\overline{64}$·52270 533 | 3610 595 | 2 | 6·00692 45 | 114 67 |
| 37 | $\overline{67}$·89595 932 | 3561 808 | 2 | ·08950 06 | 110 06 |
| 38 | $\overline{69}$·23359 169 | 3514 945 | 2 | ·17097 55 | 105 80 |
| 39 | $\overline{72}$·53607 131 | 3469 873 | 2 | ·25139 19 | 101 78 |
| 40 | $\overline{75}$·80384 908 | −3426 497 | −2 | −6·33079 00 | + 98 03 |
| 41 | $\overline{77}$·03735 897 | 3384 698 | 1 | ·40920 74 | 94 49 |
| 42 | $\overline{80}$·23701 913 | 3344 393 | 1 | ·48667 95 | 91 15 |
| 43 | $\overline{83}$·40323 277 | 3305 493 | 1 | ·56323 97 | 88 03 |
| 44 | $\overline{86}$·53638 905 | 3267 911 | 1 | ·63891 93 | 85 06 |
| 45 | $\overline{89}$·63686 391 | −3231 589 | −1 | −6·71374 80 | + 82 25 |
| 46 | $\overline{92}$·70502 071 | 3196 441 | 1 | ·78775 39 | 79 60 |
| 47 | $\overline{95}$·74121 103 | 3162 421 | 1 | ·86096 35 | 77 11 |
| 48 | $\overline{98}$·74577 519 | 3129 458 | 1 | 6·93340 18 | 74 70 |
| 49 | $\overline{101}$·71904 291 | 3097 504 | 1 | 7·00509 28 | 72 46 |
| 50 | $\overline{104}$·66133 382 | −3066 510 | −1 | −7·07605 90 | + 70 30 |
| 51 | $\overline{107}$·57295 795 | 3036 427 | 1 | ·14632 20 | 68 25 |
| 52 | $\overline{110}$·45421 622 | 3007 207 | 1 | ·21590 23 | 66 32 |
| 53 | $\overline{113}$·30540 089 | 2978 818 | 1 | ·28481 93 | 64 46 |
| 54 | $\overline{116}$·12679 593 | 2951 214 | 1 | ·35309 16 | 62 67 |
| 55 | $\overline{120}$·91867 744 | −2924 366 | −1 | −7·42073 70 | + 60 99 |
| 56 | $\overline{123}$·68131 397 | 2898 233 | 1 | ·48777 24 | 59 35 |
| 57 | $\overline{126}$·41496 690 | 2872 789 | 1 | ·55421 41 | 57 83 |
| 58 | $\overline{129}$·11989 073 | 2848 004 | 1 | ·62007 74 | 56 34 |
| 59 | $\overline{133}$·79633 336 | 2823 847 | 1 | ·68537 72 | 54 91 |
| 60 | $\overline{136}$·44453 641 | −2800 294 | −1 | −7·75012 78 | + 53 56 |
| 61 | $\overline{139}$·06473 545 | 2777 322 | 1 | ·81434 27 | 52 25 |
| 62 | $\overline{143}$·65716 025 | 2754 902 | 1 | ·87803 50 | 51 00 |
| 63 | $\overline{146}$·22203 504 | 2733 020 | −1 | 7·94121 72 | 49 78 |
| 64 | $\overline{150}$·75957 869 | 2711 648 | | 8·00390 15 | 48 65 |
| 65 | $\overline{153}$·27000 495 | −2690 771 | | −8·06609 93 | + 47 51 |
| 66 | $\overline{157}$·75352 263 | 2670 366 | | ·12782 19 | 46 45 |
| 67 | $\overline{160}$·21033 581 | 2650 419 | | ·18907 99 | 45 43 |
| 68 | $\overline{164}$·64064 399 | 2630 915 | | ·24988 36 | 44 41 |
| 69 | $\overline{167}$·04464 225 | 2611 829 | | ·31024 31 | 43 46 |
| 70 | $\overline{171}$·42252 146 | −2593 157 | | −8·37016 79 | + 42 55 |
| 71 | $\overline{175}$·77446 838 | 2574 877 | | ·42966 72 | 41 63 |
| 72 | $\overline{178}$·10066 583 | 2556 978 | | ·48875 01 | 40 80 |
| 73 | $\overline{182}$·40129 282 | 2539 449 | | ·54742 50 | 39 96 |
| 74 | $\overline{186}$·67652 467 | 2522 273 | | ·60570 03 | 39 13 |
| 75 | $\overline{190}$·92653 316 | −2505 442 | | −8·66358 42 | + 38 38 |

Interpolation formula: $f_\theta = (1-\theta)f_0 + \theta f_1 + E_0^2\delta^2_{m0} + E_1^2\delta^2_{m1} + T^4(\gamma^4_0 + \gamma^4_1)$

TABLE III—ZEROS AND TURNING-VALUES OF Ai($x$) AND Ai$'$($x$)

| $s$ | $a_s$ | Ai$'(a_s)$ | $a'_s$ | Ai$(a'_s)$ |
|---|---|---|---|---|
| 1 | − 2·33810 741 | +0·70121 082 | − 1·01879 297 | +0·53565 666 |
| 2 | 4·08794 944 | − ·80311 137 | 3·24819 758 | − ·41901 548 |
| 3 | 5·52055 983 | + ·86520 403 | 4·82009 921 | + ·38040 647 |
| 4 | 6·78670 809 | − ·91085 074 | 6·16330 736 | − ·35790 794 |
| 5 | − 7·94413 359 | +0·94733 571 | − 7·37217 726 | +0·34230 124 |
| 6 | 9·02265 085 | −0·97792 281 | 8·48848 673 | − ·33047 623 |
| 7 | 10·04017 434 | +1·00437 012 | 9·53544 905 | + ·32102 229 |
| 8 | 11·00852 430 | − ·02773 869 | 10·52766 040 | − ·31318 539 |
| 9 | 11·93601 556 | + ·04872 065 | 11·47505 663 | + ·30651 729 |
| 10 | − 12·82877 675 | − 1·06779 386 | − 12·38478 837 | −0·30073 083 |
| 11 | 13·69148 904 | + ·08530 283 | 13·26221 896 | + ·29563 148 |
| 12 | 14·52782 995 | − ·10150 457 | 14·11150 197 | − ·29108 168 |
| 13 | 15·34075 514 | + ·11659 618 | 14·93593 720 | + ·28698 071 |
| 14 | 16·13268 516 | − ·13073 231 | 15·73820 137 | − ·28325 274 |
| 15 | − 16·90563 400 | + 1·14403 667 | − 16·52050 383 | +0·27983 931 |
| 16 | 17·66130 011 | − ·15660 985 | 17·28469 505 | − ·27669 445 |
| 17 | 18·40113 260 | + ·16853 478 | 18·03234 462 | + ·27378 139 |
| 18 | 19·12638 047 | − ·17988 073 | 18·76479 844 | − ·27107 028 |
| 19 | 19·83812 989 | + ·19070 613 | 19·48322 166 | + ·26853 658 |
| 20 | − 20·53733 291 | − 1·20106 079 | − 20·18863 151 | −0·26615 987 |
| 21 | 21·22482 994 | + ·21098 751 | 20·88192 276 | + ·26392 299 |
| 22 | 21·90136 760 | − ·22052 337 | 21·56388 772 | − ·26181 141 |
| 23 | 22·56761 292 | + ·22970 070 | 22·23523 229 | + ·25981 267 |
| 24 | 23·22416 500 | − ·23854 788 | 22·89658 874 | − ·25791 608 |
| 25 | − 23·87156 446 | + 1·24708 995 | − 23·54852 630 | +0·25611 233 |
| 26 | 24·51030 124 | − ·25534 914 | 24·19155 971 | − ·25439 334 |
| 27 | 25·14082 117 | + ·26334 528 | 24·82615 643 | + ·25275 199 |
| 28 | 25·76353 140 | − ·27109 613 | 25·45274 256 | − ·25118 201 |
| 29 | 26·37880 505 | + ·27861 764 | 26·07170 794 | + ·24967 785 |
| 30 | − 26·98698 511 | − 1·28592 424 | − 26·68341 033 | −0·24823 455 |
| 31 | 27·58838 781 | + ·29302 898 | 27·28817 912 | + ·24684 770 |
| 32 | 28·18330 550 | − ·29994 375 | 27·88631 841 | − ·24551 333 |
| 33 | 28·77200 917 | + ·30667 937 | 28·47810 968 | + ·24422 787 |
| 34 | 29·35475 056 | − ·31324 575 | 29·06381 416 | − ·24298 808 |
| 35 | − 29·93176 412 | +1·31965 196 | − 29·64367 481 | +0·24179 106 |
| 36 | 30·50326 861 | − ·32590 636 | 30·21791 812 | − ·24063 412 |
| 37 | 31·06946 859 | + ·33201 664 | 30·78675 565 | + ·23951 486 |
| 38 | 31·63055 566 | − ·33798 992 | 31·35038 538 | − ·23843 104 |
| 39 | 32·18670 965 | + ·34383 277 | 31·90899 296 | + ·23738 066 |
| 40 | − 32·73809 961 | − 1·34955 130 | − 32·46275 275 | −0·23636 183 |
| 41 | 33·28488 468 | + ·35515 118 | 33·01182 878 | + ·23537 284 |
| 42 | 33·82721 495 | − ·36063 770 | 33·55637 561 | − ·23441 211 |
| 43 | 34·36523 213 | + ·36601 579 | 34·09653 909 | + ·23347 818 |
| 44 | 34·89907 025 | − ·37129 005 | 34·63245 705 | − ·23256 968 |
| 45 | − 35·42885 619 | + 1·37646 480 | − 35·16425 990 | +0·23168 536 |
| 46 | 35·95471 026 | − ·38154 407 | 35·69207 120 | − ·23082 406 |
| 47 | 36·47674 664 | + ·38653 165 | 36·21600 815 | + ·22998 469 |
| 48 | 36·99507 385 | − ·39143 111 | 36·73618 208 | − ·22916 622 |
| 49 | 37·50979 509 | + ·39624 580 | 37·25269 882 | + ·22836 772 |
| 50 | − 38·02100 868 | − 1·40097 888 | − 37·76565 910 | −0·22758 829 |

# Table IV—Bi(x) and Reduced Derivatives

| x | Bi(x) | $\tau$ | $\tau^2$ | $\tau^3$ | $\tau^4$ | $\tau^5$ | $\tau^6$ | $\tau^7$ | $\tau^8$ |
|---|---|---|---|---|---|---|---|---|---|
| | | | $0.0$ | $0.0^2$ | $0.0^3$ | $0.0^4$ | $0.0^6$ | $0.0^7$ | $0.0^9$ |
| 0·0 | +0·61492 663 | +0·04482 8836 | +0000 00000 | +010 24878 | +00 37357 | +0 00000 | +0034 | +001 | |
| 0·1 | 0·65986 169 | ·04515 1263 | 0032 99308 | 011 75022 | 00 37901 | 0 00224 | 0040 | 001 | |
| 0·2 | 0·70546 420 | ·04617 8928 | 0070 54642 | 013 29703 | 00 39658 | 0 00486 | 0047 | 001 | |
| 0·3 | 0·75248 559 | ·04800 4903 | 0112 87284 | 014 94167 | 00 42826 | 0 00788 | 0054 | 001 | |
| 0·4 | 0·80177 300 | ·05072 8168 | 0160 35460 | 016 74476 | 00 47619 | 0 01137 | 0062 | 001 | |
| 0·5 | +0·85427 704 | +0·05445 7256 | +0213 56926 | +018 77606 | +00 54280 | +0 01537 | +0072 | +001 | |
| 0·6 | 0·91106 334 | ·05931 4448 | 0273 31900 | 021 11583 | 00 63095 | 0 02000 | 0083 | 002 | |
| 0·7 | 0·97332 866 | ·06544 0592 | 0340 66503 | 023 85688 | 00 74406 | 0 02538 | 0097 | 002 | |
| 0·8 | 1·04242 217 | ·07300 0690 | 0416 96887 | 027 10713 | 00 88632 | 0 03169 | 0114 | 003 | |
| 0·9 | 1·11987 281 | ·08219 0389 | 0503 94277 | 030 99311 | 01 06288 | 0 03914 | 0135 | 003 | |
| 1·0 | +1·20742 359 | +0·09324 3593 | +0603 71180 | +035 66433 | +01 28012 | +0 04802 | +0162 | +004 | |
| 1·1 | 1·30707 425 | ·10644 1465 | 0718 89084 | 041 29884 | 01 54600 | 0 05866 | 0194 | 005 | |
| 1·2 | 1·42113 368 | ·12212 3140 | 0852 68021 | 048 11019 | 01 87037 | 0 07150 | 0235 | 006 | |
| 1·3 | 1·55228 416 | ·14069 8585 | 1008 98471 | 056 35610 | 02 26555 | 0 08708 | 0286 | 008 | |
| 1·4 | 1·70365 971 | ·16266 4116 | 1192 56180 | 066 34929 | 02 74686 | 0 10607 | 0349 | 010 | |
| 1·5 | +1·87894 150 | +0·18862 1225 | +1409 20613 | +078 47100 | +03 33335 | +0 12931 | +0428 | +013 | +0 |
| 1·6 | 2·08247 417 | ·21929 9544 | 1665 97934 | 093 18778 | 04 04880 | 0 15785 | 0527 | 016 | 0 |
| 1·7 | 2·31940 751 | ·25558 4936 | 1971 49638 | 111 07252 | 04 92283 | 0 19299 | 0649 | 020 | 1 |
| 1·8 | 2·59586 936 | ·29855 4005 | 2336 28242 | 132 83069 | 05 99237 | 0 23636 | 0802 | 024 | 1 |
| 1·9 | 2·91917 686 | ·34951 6586 | 2773 21802 | 159 33320 | 07 30357 | 0 29003 | 0994 | 031 | 1 |
| 2·0 | +3·29809 500 | +0·41006 8205 | +3298 09500 | +191 65765 | +08 91406 | +0 35656 | +1233 | +038 | +1 |
| 2·1 | 3·74315 356 | ·48215 4993 | 3930 31124 | 231 14014 | 10 89600 | 0 43921 | 1533 | 048 | 1 |
| 2·2 | 4·26703 658 | ·56815 4177 | 4693 74024 | 279 44047 | 13 33981 | 0 54207 | 1910 | 060 | 2 |
| 2·3 | 4·88506 158 | ·67097 4081 | 5617 82082 | 338 62442 | 16 35894 | 0 67031 | 2383 | 076 | 2 |
| 2·4 | 5·61577 065 | ·79417 8588 | 6738 92478 | 411 26761 | 20 09600 | 0 83047 | 2979 | 095 | 3 |
| 2·5 | +6·48166 074 | +0·94214 2332 | +8102 07592 | +500 58698 | +24 73051 | +1 03084 | +3729 | +120 | +4 |

$$\text{Bi}'(x) = 10\tau \qquad \tau^n = (h^n/n!)\,\text{Bi}^{(n)}(x) \qquad \text{Bi}(x + \theta h) = \text{Bi}(x) + \theta\tau + \theta^2\tau^2 + \theta^3\tau^3 + \ldots \qquad h = 0\cdot1$$

# Table V—Zeros and Turning-Values of Bi(x) and Bi'(x)

| s | $b_s$ | $\text{Bi}'(b_s)$ | $b_s'$ | $\text{Bi}(b_s')$ |
|---|---|---|---|---|
| 1 | − 1·17371 322 | +0·60195 789 | − 2·29443 968 | −0·45494 438 |
| 2 | 3·27109 330 | − ·76031 014 | 4·07315 509 | + ·39652 284 |
| 3 | 4·83073 784 | + ·83699 101 | 5·51239 573 | − ·36796 916 |
| 4 | 6·16985 213 | − ·88947 990 | 6·78129 445 | + ·34949 912 |
| 5 | − 7·37676 208 | +0·92998 364 | − 7·94017 869 | −0·33602 624 |
| 6 | 8·49194 885 | − ·96323 443 | 9·01958 336 | + ·32550 974 |
| 7 | 9·53819 438 | + ·99158 637 | 10·03769 633 | − ·31693 465 |
| 8 | 10·52991 351 | − 1·01638 966 | 11·00646 267 | + ·30972 594 |
| 9 | 11·47695 355 | + ·03849 429 | 11·93426 165 | − ·30352 766 |
| 10 | −12·38641 714 | − 1·05847 184 | −12·82725 831 | +0·29810 491 |
| 11 | 13·26363 952 | + ·07672 615 | 13·69015 583 | − ·29329 488 |
| 12 | 14·11275 681 | − ·09355 362 | 14·52664 576 | + ·28898 031 |
| 13 | 14·93705 741 | + ·10917 864 | 15·33969 308 | − ·28507 409 |
| 14 | 15·73921 035 | − ·12377 533 | 16·13172 478 | + ·28150 982 |
| 15 | −16·52141 955 | + 1·13748 171 | −16·90475 941 | −0·27823 579 |
| 16 | 17·28553 162 | − ·15040 911 | 17·66049 874 | + ·27521 095 |
| 17 | 18·03311 329 | + ·16264 874 | 18·40039 437 | − ·27240 223 |
| 18 | 18·76550 828 | − ·17427 625 | 19·12569 716 | + ·26978 261 |
| 19 | 19·48388 013 | + ·18535 520 | 19·83749 472 | − ·26732 974 |
| 20 | −20·18924 479 | − 1·19593 948 | −20·53674 024 | +0·26502 492 |

# TABLE IV—Bi($x$) AND REDUCED DERIVATIVES

| $x$ | Bi($-x$) | $-\tau$ | $+\tau^2$ | $-\tau^3$ | $+\tau^4$ | $-\tau^5$ | $+\tau^6$ | $-\tau^7$ | $+\tau^8$ |
|---|---|---|---|---|---|---|---|---|---|
| | | | $0.0^2$ | $0.0^3$ | $0.0^4$ | $0.0^5$ | $0.0^7$ | $0.0^8$ | $0.0^9$ |
| 0·0 | +0·61492 663 | +0·04482 8836 | −000 00000 | +10 24878 | +0 37357 | −00000 | +034 | +01 | |
| 0·1 | ·56999 904 | ·04512 1336 | 028 49995 | 08 74796 | 0 37839 | 00186 | 028 | 01 | |
| 0·2 | ·52450 903 | ·04593 8529 | 052 45090 | 07 21053 | 0 39156 | 00334 | 021 | 01 | |
| 0·3 | ·47797 784 | ·04718 8022 | 071 69668 | 05 60690 | 0 41116 | 00443 | 015 | 01 | |
| 0·4 | ·43002 094 | ·04877 3486 | 086 00419 | 03 91545 | 0 43511 | 00508 | +007 | 01 | |
| 0·5 | +0·38035 266 | +0·05059 3371 | −095 08816 | +02 12310 | +0 46123 | −00529 | −001 | +01 | |
| 0·6 | ·32879 184 | ·05254 0115 | 098 63755 | +00 22585 | 0 48715 | 00500 | 009 | 01 | |
| 0·7 | ·27526 801 | ·05449 9912 | 096 34380 | −01 77052 | 0 51037 | 00420 | 018 | 01 | |
| 0·8 | ·21982 751 | ·05635 3059 | 087 93100 | 03 84995 | 0 52823 | 00286 | 027 | 01 | |
| 0·9 | ·16263 895 | ·05797 4926 | 073 18753 | 05 98559 | 0 53802 | −00097 | 036 | 01 | |
| 1·0 | +0·10399 739 | +0·05923 7563 | −051 99869 | −08 13964 | +0 53698 | +00147 | −045 | +01 | |
| 1·1 | + ·04432 659 | ·06001 1970 | −024 37962 | 10 26342 | 0 52245 | 00443 | 053 | 01 | |
| 1·2 | − ·01582 137 | ·06017 1016 | +009 49282 | 12 29789 | 0 49193 | 00785 | 061 | 01 | |
| 1·3 | ·07576 964 | ·05959 2975 | 049 25027 | 14 17464 | 0 44325 | 01168 | 066 | 01 | |
| 1·4 | ·13472 406 | ·05816 5624 | 094 30684 | 15 81738 | 0 37469 | 01579 | 070 | +00 | |
| 1·5 | −0·19178 486 | +0·05579 0810 | +143 83865 | −17 14412 | +0 28513 | +02005 | −071 | −00 | |
| 1·6 | ·24596 320 | ·05238 9354 | 196 77056 | 18 06988 | 0 17422 | 02429 | 070 | 01 | |
| 1·7 | ·29620 266 | ·04790 6134 | 251 77226 | 18 51012 | +0 04254 | 02832 | 064 | 01 | |
| 1·8 | ·34140 583 | ·04231 5137 | 307 26525 | 18 38464 | −0 10827 | 03191 | 055 | 02 | |
| 1·9 | ·38046 588 | ·03562 4251 | 361 44258 | 17 62211 | 0 27542 | 03481 | 041 | 02 | |
| 2·0 | −0·41230 259 | +0·02787 9517 | +412 30259 | −16 16488 | −0 45484 | +03678 | −024 | −03 | |
| 2·1 | ·43590 235 | ·01916 8563 | 457 69747 | 13 97404 | 0 64123 | 03756 | −002 | 03 | |
| 2·2 | ·45036 098 | + ·00962 2919 | 495 39708 | 11 03442 | 0 82804 | 03691 | +024 | 04 | |
| 2·3 | ·45492 823 | − ·00058 1106 | 523 16747 | 07 35938 | 1 00758 | 03462 | 053 | 04 | |
| 2·4 | ·44905 228 | ·01122 3237 | 538 86273 | −02 99491 | 1 17125 | 03054 | 084 | 05 | |
| 2·5 | −0·43242 247 | −0·02204 2015 | +540 52809 | +01 97713 | −1 30978 | +02455 | +116 | −05 | |
| 2·6 | ·40500 828 | ·03273 9717 | 526 51076 | 07 43707 | 1 41360 | 01666 | 147 | 04 | |
| 2·7 | ·36709 211 | ·04298 9534 | 495 57435 | 13 22709 | 1 47329 | +00692 | 177 | 04 | |
| 2·8 | ·31929 389 | ·05244 5040 | 447 01144 | 19 15279 | 1 48007 | −00446 | 202 | 03 | |
| 2·9 | ·26258 500 | ·06075 1829 | 380 74825 | 24 98697 | 1 42641 | 01710 | 221 | 02 | |
| 3·0 | −0·19828 963 | −0·06756 1122 | +297 43444 | +30 47573 | −1 30660 | −03084 | +232 | −01 | |
| 3·1 | ·12807 165 | ·07254 4957 | 198 51106 | 35 34703 | 1 11736 | 04486 | 233 | +01 | |
| 3·2 | − ·05390 576 | ·07541 2455 | +086 24921 | 39 32155 | 0 85844 | 05860 | 223 | 02 | |
| 3·3 | + ·02196 800 | ·07592 6518 | −036 24720 | 42 12572 | 0 53304 | 07132 | 199 | 04 | |
| 3·4 | ·09710 619 | ·07392 0163 | 165 08052 | 43 50653 | −0 14827 | 08222 | 162 | 06 | |
| 3·5 | +0·16893 984 | −0·06931 1628 | −295 64472 | +43 24745 | +0 28470 | −09047 | +111 | +08 | |
| 3·6 | ·23486 631 | ·06211 7283 | 422 75935 | 41 18481 | 0 75063 | 09527 | +047 | 10 | |
| 3·7 | ·29235 261 | ·05246 1361 | 540 85233 | 37 22372 | 1 23045 | 09591 | −028 | 11 | |
| 3·8 | ·33904 647 | ·04058 1592 | 644 18829 | 31 35245 | 1 70175 | 09178 | 111 | 12 | |
| 3·9 | ·37289 058 | ·02682 9836 | 727 13663 | 23 65424 | 2 13961 | 08248 | 199 | 13 | |
| 4·0 | +0·39223 471 | −0·01166 7057 | −784 46941 | +14 31528 | +2 51767 | −06785 | −288 | +12 | |
| 4·1 | ·39593 974 | + ·00434 7872 | 811 67647 | +03 62795 | 2 80946 | 04802 | 372 | 11 | |
| 4·2 | ·38346 736 | ·02057 5691 | 805 28146 | −08 01186 | 2 98995 | −02344 | 445 | 09 | |
| 4·3 | ·35494 906 | ·03632 0468 | 763 14048 | 20 11385 | 3 03726 | +00509 | 502 | 07 | |
| 4·4 | ·31122 860 | ·05085 8932 | 684 70292 | 32 10941 | 2 93440 | 03641 | 537 | +03 | |
| 4·5 | +0·25387 266 | +0·06347 4477 | −571 21348 | −43 37465 | +2 67100 | +06903 | −545 | −01 | +1 |
| 4·6 | ·18514 576 | ·07349 4444 | 425 83524 | 53 25998 | 2 24482 | 10121 | 522 | 06 | 1 |
| 4·7 | ·10794 695 | ·08032 8926 | 253 67532 | 61 12521 | 1 66297 | 13096 | 464 | 11 | 1 |
| 4·8 | + ·02570 779 | ·08350 8976 | −061 69870 | 66 37872 | 0 94270 | 15622 | 372 | 16 | 1 |
| 4·9 | − ·05774 655 | ·08272 1903 | +141 47906 | 68 51866 | +0 11164 | 17494 | 247 | 20 | 1 |
| 5·0 | −0·13836 913 | +0·07784 1177 | +345 92284 | −67 17380 | −0 79267 | +18523 | −092 | −24 | +0 |

Bi$'(-x)=-10\tau$   $(-\tau)^n=(h^n/n!)\,\mathrm{Bi}^{(n)}(-x)$   Bi$\{-(x+\theta h)\}=\mathrm{Bi}(-x)+\theta\tau+\theta^2\tau^2+\theta^3\tau^3+\ldots$   $h=0·1$

# TABLE IV—Bi($x$) AND REDUCED DERIVATIVES

| $x$ | Bi($-x$) | $-\tau$ | $+\tau^2$ | $-\tau^3$ | $+\tau^4$ | $-\tau^5$ | $+\tau^6$ | $-\tau^7$ | $+\tau^8$ |
|---|---|---|---|---|---|---|---|---|---|
| | | | $0 \cdot 0$ | $0 \cdot 0^2$ | $0 \cdot 0^3$ | $0 \cdot 0^5$ | $0 \cdot 0^6$ | $0 \cdot 0^7$ | $0 \cdot 0^9$ |
| 5·0 | −0·13836 913 | +0·07784 1177 | +0345 92284 | −067 17380 | −00 79267 | +18523 | −0092 | −024 | 0 |
| 5·1 | ·21208 913 | ·06894 8513 | 0540 82729 | 062 14105 | 01 72395 | 18550 | +0086 | 027 | 0 |
| 5·2 | ·27502 704 | ·05634 5898 | 0715 07031 | 053 41690 | 02 62909 | 17464 | 0278 | 028 | 0 |
| 5·3 | ·32371 608 | ·04055 5694 | 0857 84760 | 041 21946 | 03 45086 | 15212 | 0472 | 027 | 0 |
| 5·4 | ·35531 708 | ·02230 7496 | 0959 35612 | 025 99870 | 04 13121 | 11816 | 0657 | 025 | 0 |
| 5·5 | −0·36781 345 | +0·00251 1158 | +1011 48700 | −008 43212 | −04 61506 | +07376 | +0818 | −021 | −1 |
| 5·6 | ·36017 223 | − ·01778 3760 | 1008 48223 | +010 59531 | 04 85445 | +02076 | 0941 | 014 | 1 |
| 5·7 | ·33245 825 | ·03744 0903 | 0947 50602 | 030 02789 | 04 81266 | −03820 | 1014 | −006 | 1 |
| 5·8 | ·28589 021 | ·05530 0203 | 0829 08161 | 048 69203 | 04 46806 | 09975 | 1026 | +003 | 1 |
| 5·9 | ·22282 969 | ·07024 7952 | 0657 34759 | 065 36332 | 03 81736 | 15995 | 0969 | 013 | 1 |
| 6·0 | −0·14669 838 | −0·08128 9879 | +0440 09513 | +078 84491 | −02 87789 | −21453 | +0838 | +024 | −1 |
| 6·1 | − ·06182 255 | ·08762 2530 | +0188 55877 | 088 05253 | 01 68869 | 25913 | 0637 | 034 | 1 |
| 6·2 | + ·02679 081 | ·08869 7896 | −0083 05151 | 092 10101 | −00 31005 | 28967 | 0371 | 042 | 1 |
| 6·3 | ·11373 701 | ·08427 6110 | 0358 27158 | 090 38553 | +01 17862 | 30263 | +0054 | 048 | 1 |
| 6·4 | ·19354 136 | ·07446 1387 | 0619 33235 | 082 65117 | 02 68259 | 29545 | −0297 | 051 | −0 |
| 6·5 | +0·26101 266 | −0·05971 7067 | −0848 29114 | +069 04370 | +04 09727 | −26681 | −0658 | +051 | +0 |
| 6·6 | ·31159 995 | ·04085 6734 | 1028 27982 | 050 13574 | 05 31507 | 21686 | 1002 | 047 | 1 |
| 6·7 | ·34172 774 | − ·01900 9878 | 1144 78792 | 026 92316 | 06 23332 | 14743 | 1302 | 038 | 1 |
| 6·8 | ·34908 418 | + ·00443 7678 | 1186 88621 | +000 78870 | 06 76267 | −06203 | 1530 | 026 | 2 |
| 6·9 | ·33283 784 | ·02792 6391 | 1148 29056 | −026 56805 | 06 83539 | +03425 | 1661 | +011 | 2 |
| 7·0 | +0·29376 207 | +0·04982 4459 | −1028 16725 | −053 23250 | +06 41285 | +13491 | −1674 | −007 | +2 |
| 7·1 | ·23425 088 | ·06854 2058 | 0831 59062 | 077 20392 | 05 49143 | 23249 | 1557 | 026 | 2 |
| 7·2 | ·15821 739 | ·08265 0634 | 0569 58260 | 096 54380 | 04 10625 | 31908 | 1307 | 045 | 2 |
| 7·3 | + ·07087 411 | ·09099 8427 | −0258 69052 | 109 53352 | 02 33202 | 38686 | 0933 | 062 | 2 |
| 7·4 | − ·02159 652 | ·09281 2809 | +0079 90712 | 114 82907 | +00 28068 | 42886 | −0452 | 075 | 1 |
| 7·5 | −0·11246 349 | +0·08778 0228 | +0421 73807 | −111 59968 | −01 90436 | +43959 | +0104 | −083 | +1 |
| 7·6 | ·19493 376 | ·07609 5509 | 0740 74827 | 099 63654 | 04 05728 | 41566 | 0696 | 085 | 0 |
| 7·7 | ·26267 007 | ·05847 4045 | 1011 27976 | 079 41953 | 06 00176 | 35633 | 1276 | 080 | −1 |
| 7·8 | ·31030 057 | ·03612 2930 | 1210 17221 | 052 13149 | 07 56509 | 26382 | 1793 | 067 | 2 |
| 7·9 | ·33387 856 | + ·01067 0215 | 1318 82032 | −019 61376 | 08 59332 | 14342 | 2198 | 047 | 3 |
| 8·0 | −0·33125 158 | −0·01594 5050 | +1325 00632 | +015 73921 | −08 96625 | +00329 | +2443 | −022 | −3 |
| 8·1 | ·30230 331 | ·04161 5664 | 1224 32840 | 051 14276 | 08 61101 | −14591 | 2495 | +008 | 4 |
| 8·2 | ·24904 019 | ·06423 2293 | 1021 06478 | 083 63346 | 07 51255 | 29184 | 2332 | 039 | 4 |
| 8·3 | ·17550 556 | ·08186 0044 | 0728 34809 | 110 31463 | 05 71991 | 42139 | 1950 | 070 | 4 |
| 8·4 | − ·08751 798 | ·09291 0958 | +0367 57552 | 128 61671 | 03 34729 | 52181 | 1366 | 096 | 3 |
| 8·5 | +0·00775 444 | −0·09629 6917 | −0032 95635 | +136 54987 | −00 56903 | −58198 | +0616 | +116 | −2 |
| 8·6 | ·10235 647 | ·09154 7918 | 0440 13282 | 132 92462 | +02 39139 | 59358 | −0242 | 127 | −1 |
| 8·7 | ·18820 363 | ·07888 2623 | 0818 68577 | 117 51653 | 05 27812 | 55213 | 1139 | 127 | +1 |
| 8·8 | ·25778 240 | ·05922 1371 | 1134 24257 | 091 15438 | 07 82427 | 45779 | 1991 | 115 | 2 |
| 8·9 | ·30483 241 | ·03413 6475 | 1356 50424 | 055 71631 | 09 77627 | 31576 | 2715 | 090 | 4 |
| 9·0 | +0·32494 732 | −0·00574 0051 | −1462 26296 | +014 02587 | +10 91914 | −13623 | −3229 | +055 | +5 |
| 9·1 | ·31603 471 | + ·02348 4379 | 1437 95794 | −030 35073 | 11 10022 | +06620 | 3468 | +012 | 6 |
| 9·2 | ·27858 425 | ·05089 4402 | 1281 48757 | 073 39501 | 10 24886 | 27354 | 3388 | −036 | 6 |
| 9·3 | ·21570 835 | ·07392 8028 | 1003 04381 | 110 99330 | 08 38966 | 46597 | 2971 | 083 | 6 |
| 9·4 | ·13293 876 | ·09034 8537 | 0624 81218 | 139 33040 | 05 64727 | 62361 | 2234 | 126 | 5 |
| 9·5 | +0·03778 543 | +0·09847 1407 | −0179 48080 | −155 28330 | +02 24148 | +72862 | −1227 | −159 | +3 |
| 9·6 | − ·06091 293 | ·09734 9918 | +0292 38205 | 156 77508 | −01 52781 | 76714 | −0034 | 179 | +1 |
| 9·7 | ·15379 421 | ·08689 8388 | 0745 90191 | 143 04896 | 05 30522 | 73108 | +1239 | 181 | −1 |
| 9·8 | ·23186 331 | ·06793 6774 | 1136 13021 | 114 82779 | 08 71226 | 61946 | 2463 | 165 | 3 |
| 9·9 | ·28738 356 | ·04214 7209 | 1422 54861 | 074 33262 | 11 38480 | 43907 | 3509 | 131 | 5 |
| 10·0 | −0·31467 983 | +0·01194 1411 | +1573 39915 | −025 14702 | −13 01215 | +20441 | +4254 | −080 | −7 |

Bi′($-x$) = −10$\tau$    $(-\tau)^n = (h^n/n!)$ Bi$^{(n)}(-x)$    Bi$\{-(x+\theta h)\}$ = Bi($-x$) + $\theta\tau + \theta^2\tau^2 + \theta^3\tau^3 + \ldots$    $h = 0\cdot1$

## Table VI—$\log_{10}\mathrm{Bi}(x)$ and $\mathrm{Bi}'(x)/\mathrm{Bi}(x)$

| $x$ | $\log_{10}\mathrm{Bi}(x)$ | $\delta^2_m$ | $\gamma^4$ | $\mathrm{Bi}'(x)/\mathrm{Bi}(x)$ | $\delta^2_m$ | $\gamma^4$ | $x$ | $\log_{10}\mathrm{Bi}(x)$ | $\delta^2_m$ | $\mathrm{Bi}'(x)/\mathrm{Bi}(x)$ | $\delta^2$ |
|---|---|---|---|---|---|---|---|---|---|---|---|
| 0·0 | $\bar{1}$·78882 330 | − 229 451 | − 13 | 0·72901 11 | + 1764 78 | + 10 | 5·0 | 2·81808 862 | + 102 248 | 2·18278 57 | − 27 86 |
| 0·1 | ·81945 292 | 158 903 | 10 | ·68425 34 | 1497 47 | 6 | 5·1 | 2·91339 506 | 101 062 | ·20619 28 | 26 82 |
| 0·2 | ·84847 498 | 98 450 | 8 | ·65458 93 | 1295 04 | 4 | 5·2 | 3·00971 221 | 99 918 | ·22933 17 | 25 86 |
| 0·3 | ·87649 818 | − 45 823 | 6 | ·63795 11 | 1133 92 | 3 | 5·3 | ·10702 861 | 98 816 | ·25221 20 | 24 95 |
| 0·4 | ·90405 143 | + 416 | 5 | ·63269 99 | 998 81 | 2 | 5·4 | ·20533 325 | 97 752 | ·27484 28 | 24 08 |
| 0·5 | ·93159 874 | + 41 162 | − 5 | 0·63746 60 | + 879 68 | + 1 | 5·5 | ·30461 546 | + 96 723 | 2·29723 28 | − 23 30 |
| 0·6 | ·95954 857 | 76 961 | 5 | ·65104 64 | 770 09 | 1 | 5·6 | ·40486 498 | 95 729 | ·31938 98 | 22 54 |
| 0·7 | $\bar{1}$·98825 951 | 108 133 | 4 | ·67233 81 | 666 21 | 0 | 5·7 | ·50607 184 | 94 766 | ·34132 14 | 21 81 |
| 0·8 | 0·01804 364 | 134 880 | 4 | ·70029 87 | 566 08 | 0 | 5·8 | ·60822 641 | 93 833 | ·36303 49 | 21 15 |
| 0·9 | ·04916 870 | 157 349 | 4 | ·73392 61 | 469 21 | 0 | 5·9 | ·71131 938 | 92 929 | ·38453 69 | 20 52 |
| 1·0 | ·08185 966 | + 175 690 | − 4 | 0·77225 25 | + 376 11 | + 0 | 6·0 | ·81534 168 | + 92 052 | 2·40583 37 | − 19 89 |
| 1·1 | ·11630 026 | 190 088 | 4 | ·81434 90 | 287 89 | 1 | 6·1 | 3·92028 456 | 91 201 | ·42693 16 | 19 33 |
| 1·2 | ·15263 493 | 200 787 | 3 | ·85933 60 | 205 98 | 1 | 6·2 | 4·02613 949 | 90 373 | ·44783 62 | 18 78 |
| 1·3 | ·19097 123 | 208 091 | 3 | ·90639 71 | 131 78 | 1 | 6·3 | ·13289 819 | 89 569 | ·46855 30 | 18 27 |
| 1·4 | ·23138 285 | 212 362 | 3 | 0·95479 23 | 66 44 | 1 | 6·4 | ·24055 263 | 88 787 | ·48908 71 | 17 76 |
| 1·5 | ·27391 326 | + 214 002 | − 2 | 1·00386 96 | + 10 70 | + 1 | 6·5 | ·34909 498 | + 88 026 | 2·50944 36 | − 17 29 |
| 1·6 | ·31857 962 | 213 435 | 2 | ·05307 21 | − 35 17 | 1 | 6·6 | ·45851 763 | 87 285 | ·52962 72 | 16 85 |
| 1·7 | ·36537 706 | 211 087 | 1 | ·10194 06 | 71 40 | 1 | 6·7 | ·56881 317 | 86 563 | ·54964 23 | 16 40 |
| 1·8 | ·41428 283 | 207 364 | 1 | ·15011 18 | 98 58 | 1 | 6·8 | ·67997 438 | 85 860 | ·56949 34 | 16 01 |
| 1·9 | ·46526 041 | 202 641 | − 1 | ·19731 21 | 117 66 | 1 | 6·9 | ·79199 422 | 85 174 | ·58918 44 | 15 59 |
| 2·0 | ·51826 316 | + 197 246 | | 1·24334 87 | − 129 75 | + 1 | 7·0 | 4·90486 582 | + 84 505 | 2·60871 95 | − 15 24 |
| 2·1 | ·57323 764 | 191 455 | | ·28809 83 | 136 06 | | 7·1 | 5·01858 251 | 83 851 | ·62810 22 | 14 85 |
| 2·2 | ·63012 637 | 185 494 | | ·33149 59 | 137 76 | | 7·2 | ·13313 773 | 83 214 | ·64733 64 | 14 52 |
| 2·3 | ·68887 004 | 179 539 | | ·37352 23 | 135 98 | | 7·3 | ·24852 512 | 82 591 | ·66642 54 | 14 18 |
| 2·4 | ·74940 936 | 173 720 | | ·41419 34 | 131 68 | | 7·4 | ·36473 844 | 81 982 | ·68537 26 | 13 86 |
| 2·5 | ·81168 630 | + 168 126 | | 1·45355 08 | − 125 70 | | 7·5 | ·48177 161 | + 81 387 | 2·70418 12 | − 13 54 |
| 2·6 | ·87564 501 | 162 816 | | ·49165 31 | 118 70 | | 7·6 | ·59961 868 | 80 805 | ·72285 44 | 13 26 |
| 2·7 | 0·94123 246 | 157 822 | | ·52856 92 | 111 21 | | 7·7 | ·71827 382 | 80 236 | ·74139 50 | 12 96 |
| 2·8 | 1·00839 875 | 153 158 | | ·56437 35 | 103 60 | | 7·8 | ·83773 135 | 79 679 | ·75980 60 | 12 69 |
| 2·9 | ·07709 722 | 148 821 | | ·59914 14 | 96 17 | | 7·9 | 5·95798 569 | 79 134 | ·77809 01 | 12 42 |
| 3·0 | ·14728 448 | + 144 800 | | 1·63294 71 | − 89 08 | | 8·0 | 6·07903 139 | + 78 600 | 2·79625 00 | − 12 17 |
| 3·1 | ·21892 029 | 141 077 | | ·66586 11 | 82 45 | | 8·1 | ·20086 311 | 78 077 | ·81428 82 | 11 92 |
| 3·2 | ·29196 739 | 137 631 | | ·69794 97 | 76 33 | | 8·2 | ·32347 563 | 77 565 | ·83220 72 | 11 68 |
| 3·3 | ·36639 126 | 134 439 | | ·72927 39 | 70 75 | | 8·3 | ·44686 381 | 77 063 | ·85000 94 | 11 44 |
| 3·4 | ·44215 996 | 131 478 | | ·75988 97 | 65 70 | | 8·4 | ·57102 265 | 76 571 | ·86769 72 | 11 23 |
| 3·5 | ·51924 382 | + 128 726 | | 1·78984 76 | − 61 14 | | 8·5 | ·69594 720 | + 76 088 | 2·88527 27 | − 11 00 |
| 3·6 | ·59761 528 | 126 161 | | ·81919 33 | 57 04 | | 8·6 | ·82163 266 | 75 615 | ·90273 82 | 10 81 |
| 3·7 | ·67724 867 | 123 765 | | ·84796 77 | 53 36 | | 8·7 | 6·94807 427 | 75 150 | ·92009 56 | 10 59 |
| 3·8 | ·75811 999 | 121 521 | | ·87620 79 | 50 06 | | 8·8 | 7·07526 741 | 74 694 | ·93734 71 | 10 40 |
| 3·9 | ·84020 677 | 119 413 | | ·90394 70 | 47 09 | | 8·9 | ·20320 750 | 74 247 | ·95449 46 | 10 21 |
| 4·0 | 1·92348 790 | + 117 427 | | 1·93121 45 | − 44 41 | | 9·0 | ·33189 007 | + 73 807 | 2·97154 00 | − 10 02 |
| 4·1 | 2·00794 350 | 115 551 | | ·95803 75 | 42 00 | | 9·1 | ·46131 074 | 73 376 | 2·98848 52 | 9 85 |
| 4·2 | ·09355 479 | 113 775 | | 1·98444 00 | 39 82 | | 9·2 | ·59146 517 | 72 952 | 3·00533 19 | 9 68 |
| 4·3 | ·18030 400 | 112 090 | | 2·01044 40 | 37 83 | | 9·3 | ·72234 915 | 72 536 | ·02208 18 | 9 50 |
| 4·4 | ·26817 426 | 110 486 | | ·03606 93 | 36 02 | | 9·4 | ·85395 849 | 72 127 | ·03873 67 | 9 34 |
| 4·5 | ·35714 952 | + 108 958 | | 2·06133 41 | − 34 37 | | 9·5 | 7·98628 911 | + 71 724 | 3·05529 82 | − 9 19 |
| 4·6 | ·44721 448 | 107 499 | | ·08625 49 | 32 85 | | 9·6 | 8·11933 699 | 71 329 | ·07176 78 | 9 02 |
| 4·7 | ·53835 456 | 106 103 | | ·11084 70 | 31 45 | | 9·7 | ·25309 817 | 70 940 | ·08814 72 | 8 89 |
| 4·8 | ·63055 577 | 104 765 | | ·13512 44 | 30 16 | | 9·8 | ·38756 876 | 70 558 | ·10443 77 | 8 73 |
| 4·9 | ·72380 474 | 103 482 | | ·15910 00 | 28 96 | | 9·9 | ·52274 494 | 70 181 | ·12064 09 | 8 59 |
| 5·0 | 2·81808 862 | + 102 248 | | 2·18278 57 | − 27 85 | | 10·0 | 8·65862 294 | + 69 811 | 3·13675 82 | − 8 45 |

Interpolation formulae: 
$$\begin{cases} \text{When } |\gamma_1^4 - \gamma_0^4| > 1 & \text{use } f_\theta = (1-\theta)f_0 + \theta f_1 + E_0^2\delta_{m0}^2 + E_1^2\delta_{m1}^2 + M_0^4\gamma_0^4 + M_1^4\gamma_1^4 \\ \text{When } |\gamma_1^4 - \gamma_0^4| \leqslant 1 & \text{use } f_\theta = (1-\theta)f_0 + \theta f_1 + E_0^2\delta_{m0}^2 + E_1^2\delta_{m1}^2 + T^4(\gamma_0^4 + \gamma_1^4) \end{cases}$$

*Definitions:*

$$\text{Ai}(x) = F(x)\sin\chi(x) \qquad\qquad \text{Ai}'(x) = G(x)\sin\psi(x)$$
$$\text{Bi}(x) = F(x)\cos\chi(x) \qquad\qquad \text{Bi}'(x) = G(x)\cos\psi(x)$$

*Asymptotic expansions:*

$$\{F(-x)\}^2 \sim \frac{1}{\pi x^{1/2}}\left(1 - \frac{1\cdot3\cdot5}{1!\,96}\frac{1}{x^3} + \frac{1\cdot3\cdot5\cdot7\cdot9\cdot11}{2!\,96^2}\frac{1}{x^6} - \frac{1\cdot3\cdot5\cdot7\cdot9\cdot11\cdot13\cdot15\cdot17}{3!\,96^3}\frac{1}{x^9} + \cdots\right)$$

$$\{G(-x)\}^2 \sim \frac{1}{\pi}x^{1/2}\left(1 + \frac{1\cdot3}{1!\,96}\frac{7}{x^3} - \frac{1\cdot3\cdot5\cdot7\cdot9}{2!\,96^2}\frac{13}{x^6} + \frac{1\cdot3\cdot5\cdot7\cdot9\cdot11\cdot13\cdot15}{3!\,96^3}\frac{19}{x^9} - \cdots\right)$$

and

$$\chi(-x) - \tfrac14\pi \sim \tfrac23 x^{3/2}\left(1 - \frac{5}{32}\frac{1}{x^3} + \frac{1105}{6144}\frac{1}{x^6} - \frac{82825}{65536}\frac{1}{x^9} + \frac{12820\,31525}{587\,20256}\frac{1}{x^{12}} - \cdots\right)$$

$$\psi(-x) + \tfrac14\pi \sim \tfrac23 x^{3/2}\left(1 + \frac{7}{32}\frac{1}{x^3} - \frac{1463}{6144}\frac{1}{x^6} + \frac{4\,95271}{3\,27680}\frac{1}{x^9} - \frac{2065\,30429}{83\,88608}\frac{1}{x^{12}} + \cdots\right)$$

*Zeros:*

If $a_s$, $b_s$, $a_s'$, $b_s'$ denote respectively the $s$th zeros of $\text{Ai}(x)$, $\text{Bi}(x)$, $\text{Ai}'(x)$, $\text{Bi}'(x)$, and if

$$\lambda = \tfrac38\pi(4s-1) \qquad\qquad \mu = \tfrac38\pi(4s-3)$$

then

$$a_s \sim -\lambda^{2/3}\left(1 + \frac{5}{48}\frac{1}{\lambda^2} - \frac{5}{36}\frac{1}{\lambda^4} + \frac{77125}{82944}\frac{1}{\lambda^6} - \frac{1080\,56875}{69\,67296}\frac{1}{\lambda^8} + \frac{16\,23755\,96875}{3344\,30208}\frac{1}{\lambda^{10}} - \cdots\right)$$

$$a_s' \sim -\mu^{2/3}\left(1 - \frac{7}{48}\frac{1}{\mu^2} + \frac{35}{288}\frac{1}{\mu^4} - \frac{1\,81223}{2\,07360}\frac{1}{\mu^6} + \frac{186\,83371}{12\,44160}\frac{1}{\mu^8} - \frac{9\,11458\,84361}{1911\,02976}\frac{1}{\mu^{10}} + \cdots\right)$$

With $\lambda = \tfrac38\pi(4s-3)$, $\mu = \tfrac38\pi(4s-1)$ these formulae give $b_s$, $b_s'$ respectively.

$$\chi(a_s) = \psi(a_{s+1}') = s\pi \qquad\qquad \chi(b_s) = \psi(b_s') = (s-\tfrac12)\pi$$

If $y$ is any solution of the differential equation $y'' = xy$, if $k$, $k'$ are approximations to zeros $c$, $c'$ of $y$, $y'$ respectively, and if

$$u = y(k)/y'(k) \qquad\qquad v = y'(k')/k'^2 y(k')$$

then

$$c = k - u - 2k\frac{u^3}{3!} + 2\frac{u^4}{4!} - 24k^2\frac{u^5}{5!} + 88k\frac{u^6}{6!} - (88 + 720k^3)\frac{u^7}{7!} + 5856k^2\frac{u^8}{8!} - (16640k + 40320k^4)\frac{u^9}{9!} + \cdots$$

$$c' = k'\left\{1 - v - \frac{v^2}{2!} - (3 + 2k'^3)\frac{v^3}{3!} - (15 + 10k'^3)\frac{v^4}{4!} - (105 + 76k'^3 + 24k'^6)\frac{v^5}{5!} - (945 + 756k'^3 + 272k'^6)\frac{v^6}{6!} - \cdots\right\}$$

*Maxima and minima:*

The maxima and minima of $\text{Ai}'(x)$, $\text{Bi}'(x)$, $\text{Ai}(x)$, $\text{Bi}(x)$ corresponding to the zeros $a_s$, $b_s$, $a_s'$, $b_s'$ are given by

$$\text{Ai}'(a_s) = (-1)^{s-1}/\pi F(a_s) \qquad\qquad \text{Ai}(a_s') = (-1)^{s-1}/\pi G(a_s')$$
$$\text{Bi}'(b_s) = (-1)^{s-1}/\pi F(b_s) \qquad\qquad \text{Bi}(b_s') = (-1)^{s}/\pi G(b_s')$$

The maxima and minima of $y'$, $y$ corresponding to the zeros $c$, $c'$ of the previous section are given by

$$y'(c) = y'(k)\left\{1 - k\frac{u^2}{2!} + \frac{u^3}{3!} - 3k^2\frac{u^4}{4!} + 14k\frac{u^5}{5!} - (14 + 45k^3)\frac{u^6}{6!} + 471k^2\frac{u^7}{7!} - (1432k + 1575k^4)\frac{u^8}{8!} + \cdots\right\}$$

$$y(c') = y(k')\left\{1 - k'^3\frac{v^2}{2!} - k'^3\frac{v^3}{3!} - (3k'^3 + 3k'^6)\frac{v^4}{4!} - (15k'^3 + 14k'^6)\frac{v^5}{5!} - (105k'^3 + 101k'^6 + 45k'^9)\frac{v^6}{6!} - \cdots\right\}$$

## Table VII—Auxiliary Functions

| $x$ | $F(x)$ | $\delta^2_m$ | $\gamma^4$ | $\chi(x)$ | $\delta^2_m$ | $\gamma^4$ |
|---|---|---|---|---|---|---|
| 0·0 | 0·71005 61 | + 282 39 | + 1 | + 30·00000 0° | +26382 5 | − 1 |
| 0·1 | 0·73742 26 | 325 69 | 1 | 26·51447 9 | 26322 6 | 1 |
| 0·2 | 0·76805 95 | 376 33 | 1 | 23·29193 4 | 26126 9 | 2 |
| 0·3 | 0·80247 59 | 435 78 | 1 | 20·33037 6 | 25778 0 | 2 |
| 0·4 | 0·84126 90 | 505 49 | 1 | 17·62629 0 | 25261 9 | 2 |
| 0·5 | 0·88513 91 | + 587 22 | + 1 | + 15·17449 3 | +24566 6 | − 2 |
| 0·6 | 0·93490 77 | 683 29 | 2 | 12·96802 4 | 23687 8 | 2 |
| 0·7 | 0·99153 98 | 795 99 | 2 | 10·99809 8 | 22627 2 | 2 |
| 0·8 | 1·05616 84 | 928 62 | 2 | 9·25412 7 | 21394 4 | 2 |
| 0·9 | 1·13012 60 | 1084 50 | 3 | 7·72381 7 | 20008 1 | 1 |
| 1·0 | 1·21497 97 | + 1268 20 | + 3 | + 6·39335 3 | +18494 2 | − 1 |
| 1·1 | 1·31257 57 | 1484 69 | 4 | 5·24766 0 | 16887 5 | − 1 |
| 1·2 | 1·42509 07 | 1740 43 | 5 | 4·27074 0 | 15225 3 | 0 |
| 1·3 | 1·55509 60 | 2042 94 | 6 | 3·44604 7 | 13549 0 | 0 |
| 1·4 | 1·70563 38 | 2401 53 | 7 | 2·75689 2 | 11898 7 | + 1 |
| 1·5 | 1·88031 09 | + 2827 58 | + 8 | + 2·18684 1 | +10311 8 | + 1 |
| 1·6 | 2·08341 30 | 3334 79 | 10 | 1·72008 1 | 8818 7 | 1 |
| 1·7 | 2·32004 36 | 3940 28 | 12 | 1·34172 7 | 7444 4 | 1 |
| 1·8 | 2·59629 55 | 4664 61 | 14 | 1·03806 5 | 6204 6 | 1 |
| 1·9 | 2·91945 91 | 5533 43 | 17 | 0·79671 1 | 5107 0 | 1 |
| 2·0 | 3·29827 99 | + 6577 90 | +21 | + 0·60669 3 | + 4153 8 | + 1 |
| 2·1 | 3·74327 34 | 7836 56 | 26 | 0·45846 9 | 3339 6 | 1 |
| 2·2 | 4·26711 34 | 9356 87 | 32 | 0·34388 0 | 2655 4 | 1 |
| 2·3 | 4·88511 04 | 11197 24 | 39 | 0·25606 1 | 2088 7 | 1 |
| 2·4 | 5·61580 13 | 13430 17 | 48 | 0·18932 1 | 1626 6 | 1 |
| 2·5 | 6·48167 98 | +16145 11 | +59 | + 0·13901 2 | + 1254 2 | + 1 |

| $x$ | $G(x)$ | $\delta^2_m$ | $\gamma^4$ | $\psi(x)$ | $\delta^2_m$ | $\gamma^4$ |
|---|---|---|---|---|---|---|
| 0·0 | 0·51763 88 | + 352 80 | + 2 | − 30·00000 0° | +68345 7 | − 28 |
| 0·1 | 0·51959 57 | 467 41 | 2 | 29·66093 6 | 66874 9 | 39 |
| 0·2 | 0·52626 79 | 604 35 | 2 | 28·66019 0 | 61567 8 | 45 |
| 0·3 | 0·53902 12 | 761 66 | 2 | 27·05201 2 | 51810 5 | 43 |
| 0·4 | 0·55942 06 | 934 99 | 1 | 24·93341 8 | 37898 1 | 28 |
| 0·5 | 0·58918 93 | + 1118 90 | + 1 | − 22·44081 1 | +21307 9 | − 3 |
| 0·6 | 0·63015 98 | 1309 86 | 1 | 19·73567 8 | + 4419 5 | +23 |
| 0·7 | 0·68424 24 | 1508 26 | 1 | 16·98232 1 | − 10298 4 | 40 |
| 0·8 | 0·75343 20 | 1720 03 | 2 | 14·32479 2 | 21155 9 | 45 |
| 0·9 | 0·83986 51 | 1955 35 | 4 | 11·87073 0 | 27639 3 | 39 |
| 1·0 | 0·94592 00 | + 2227 86 | + 5 | − 9·68586 5 | − 30223 4 | +29 |
| 1·1 | 1·07434 93 | 2552 46 | 7 | 7·79787 3 | 29895 1 | 18 |
| 1·2 | 1·22842 91 | 2945 52 | 9 | 6·20543 4 | 27715 6 | 9 |
| 1·3 | 1·41212 23 | 3424 62 | 11 | 4·88838 3 | 24569 2 | + 3 |
| 1·4 | 1·63025 64 | 4009 62 | 13 | 3·81641 9 | 21088 5 | − 1 |
| 1·5 | 1·88872 44 | + 4723 91 | + 16 | − 2·95546 2 | − 17669 9 | − 3 |
| 1·6 | 2·19472 03 | 5595 27 | 19 | 2·27172 7 | 14532 6 | 4 |
| 1·7 | 2·55702 03 | 6657 77 | 23 | 1·73402 3 | 11776 8 | 4 |
| 1·8 | 2·98632 63 | 7953 27 | 28 | 1·31483 5 | 9426 6 | 4 |
| 1·9 | 3·49568 83 | 9533 44 | 35 | 0·99063 3 | 7467 1 | 4 |
| 2·0 | 4·10102 57 | +11462 33 | + 43 | − 0·74175 1 | − 5859 7 | − 3 |
| 2·1 | 4·82177 37 | 13819 55 | 52 | 0·55203 2 | 4559 8 | 3 |
| 2·2 | 5·68168 61 | 16703 91 | 65 | 0·40839 0 | 3519 9 | 2 |
| 2·3 | 6·70983 30 | 20238 65 | 80 | 0·30034 7 | 2697 5 | 2 |
| 2·4 | 7·94184 42 | 24577 40 | 99 | 0·21960 4 | 2051 8 | 1 |
| 2·5 | 9·42145 99 | +29912 07 | + 1 23 | − 0·15964 3 | − 1549 7 | − 1 |

Interpolation formulae:
$$\text{When } |\gamma_1^4 - \gamma_0^4| > 1 \text{ use } f_\theta = (1-\theta)f_0 + \theta f_1 + E_0^2\delta_{m0}^2 + E_1^2\delta_{m1}^2 + M_0^4\gamma_0^4 + M_1^4\gamma_1^4$$
$$\text{When } |\gamma_1^4 - \gamma_0^4| \leqslant 1 \text{ use } f_\theta = (1-\theta)f_0 + \theta f_1 + E_0^2\delta_{m0}^2 + E_1^2\delta_{m1}^2 + T^4(\gamma_0^4 + \gamma_1^4)$$

TABLE VII—AUXILIARY FUNCTIONS

| $x$ | $F(-x)$ | $\delta^2_m$ | $\gamma^4$ | $\chi(-x)$ | $\delta^2_m$ | $\gamma^4$ | $G(-x)$ | $\delta^2_m$ | $\gamma^4$ | $\psi(-x)$ | $\delta^2_m$ | $\gamma^4$ |
|---|---|---|---|---|---|---|---|---|---|---|---|---|
| | 0. | + | + | 0 | + | − | 0. | | + | − 0 | + | |
| 0·0 | 71005 611 | 282 390 | 6 | 30.00000 0 | 26382 5 | 1 | 51763 881 | +352 805 | 22 | 30.00000 0 | 68345 7 | −28 |
| 0·1 | 68552 507 | 245 417 | 5 | 33.74913 0 | 26325 1 | 1 | 51925 039 | 260 156 | 20 | 29.66075 6 | 67017 6 | 17 |
| 0·2 | 66345 804 | 213 796 | 5 | 37.76132 8 | 26168 6 | 1 | 52350 000 | 187 301 | 17 | 28.65455 0 | 63937 4 | 9 |
| 0·3 | 64353 733 | 186 722 | 4 | 42.03506 1 | 25930 5 | 1 | 52965 356 | 131 246 | 14 | 27.01061 7 | 59955 9 | − 3 |
| 0·4 | 62549 094 | 163 510 | 3 | 46.57797 8 | 25627 2 | 1 | 53714 470 | 88 840 | 11 | 24.76768 4 | 55665 1 | + 1 |
| 0·5 | 60908 568 | 143 576 | 3 | 51.35707 3 | 25272 9 | | 54554 400 | + 57 176 | 8 | − 21.96798 0 | 51435 7 | + 3 |
| 0·6 | 59412 131 | 126 431 | 2 | 56.39882 7 | 24881 0 | | 55453 031 | 33 808 | 6 | 18.65342 6 | 47471 5 | 4 |
| 0·7 | 58042 561 | 111 655 | 2 | 61.68934 0 | 24461 7 | | 56386 625 | 16 724 | 5 | 14.86349 4 | 43865 9 | 4 |
| 0·8 | 56785 017 | 98 896 | 2 | 67.22443 6 | 24024 4 | | 57337 813 | + 4 380 | 4 | 10.63419 2 | 40645 8 | 4 |
| 0·9 | 55626 685 | 87 855 | 1 | 72.99975 6 | 23576 8 | | 58294 027 | − 4 447 | 3 | 5.99774 2 | 37800 4 | 4 |
| 1·0 | 54556 477 | 78 276 | 1 | 79.01083 5 | 23124 5 | | 59246 276 | − 10 647 | 2 | − 0.98265 4 | 35299 3 | + 3 |
| 1·1 | 53564 775 | 69 948 | 1 | 85.25316 0 | 22673 1 | | 60188 233 | 14 916 | 1 | + 4.38599 6 | 33107 6 | 3 |
| 1·2 | 52643 218 | 62 691 | 1 | 91.72222 4 | 22226 0 | | 61115 536 | 17 758 | 1 | 10.08621 7 | 31184 6 | 2 |
| 1·3 | 51784 520 | 56 348 | 1 | 98.41356 1 | 21786 1 | | 62025 273 | 19 556 | 1 | 16.09871 5 | 29496 0 | 2 |
| 1·4 | 50982 315 | 50 793 | 1 | 105.32277 6 | 21355 8 | | 62915 594 | 20 593 | 1 | 22.40654 3 | 28008 4 | 2 |
| 1·5 | 50231 027 | 45 911 | 1 | 112.44556 9 | 20936 5 | | 63785 424 | − 21 075 | | + 28.99477 2 | 26693 0 | + 1 |
| 1·6 | 49525 758 | 41 617 | 1 | 119.77774 9 | 20529 1 | | 64634 251 | 21 166 | | 35.85020 3 | 25525 4 | 1 |
| 1·7 | 48862 198 | 37 822 | | 127.31524 4 | 20135 0 | | 65461 965 | 20 968 | | 42.96112 1 | 24484 6 | 1 |
| 1·8 | 48236 540 | 34 464 | | 135.05411 3 | 19753 9 | | 66268 746 | 20 580 | | 50.31708 4 | 23552 0 | 1 |
| 1·9 | 47645 415 | 31 483 | | 142.99054 5 | 19386 0 | | 67054 972 | 20 055 | | 57.90874 0 | 22713 5 | 1 |
| 2·0 | 47085 833 | 28 828 | | 151.12086 2 | 19031 7 | | 67821 159 | − 19 443 | | + 65.72767 9 | 21955 3 | + 1 |
| 2·1 | 46555 133 | 26 467 | | 159.44152 0 | 18690 4 | | 68567 912 | 18 783 | | 73.76630 0 | 21267 2 | |
| 2·2 | 46050 945 | 24 347 | | 167.94910 6 | 18362 1 | | 69295 888 | 18 087 | | 82.01770 5 | 20640 1 | + 1 |
| 2·3 | 45571 145 | 22 453 | | 176.64033 6 | 18046 3 | | 70005 778 | 17 388 | | 90.47560 8 | 20065 8 | |
| 2·4 | 45113 833 | 20 748 | | 185.51205 1 | 17742 4 | | 70698 280 | 16 689 | | 99.13425 5 | 19538 3 | |
| 2·5 | 44677 300 | 19 212 | | 194.56121 2 | 17450 5 | | 71374 091 | − 16 003 | | + 107.98836 0 | 19051 4 | |
| 2·6 | 44260 007 | 17 827 | | 203.78489 8 | 17169 5 | | 72033 896 | 15 334 | | 117.03304 6 | 18601 0 | |
| 2·7 | 43860 565 | 16 571 | | 213.18029 9 | 16899 3 | | 72678 363 | 14 687 | | 126.26380 1 | 18182 7 | |
| 2·8 | 43477 716 | 15 435 | | 222.74471 2 | 16639 7 | | 73308 138 | 14 067 | | 135.67643 5 | 17792 7 | |
| 2·9 | 43110 321 | 14 401 | | 232.47553 9 | 16389 2 | | 73923 841 | 13 472 | | 145.26704 3 | 17428 2 | |
| 3·0 | 42757 344 | 13 460 | | 242.37027 6 | 16148 7 | | 74526 067 | − 12 904 | | + 155.03197 6 | 17086 9 | |
| 3·1 | 42417 842 | 12 600 | | 252.42651 6 | 15916 8 | | 75115 384 | 12 365 | | 164.96781 6 | 16766 1 | |
| 3·2 | 42090 954 | 11 814 | | 262.64193 9 | 15693 2 | | 75692 331 | 11 851 | | 175.07135 1 | 16463 8 | |
| 3·3 | 41775 893 | 11 098 | | 273.01430 9 | 15478 0 | | 76257 422 | 11 362 | | 185.33955 5 | 16178 6 | |
| 3·4 | 41471 940 | 10 433 | | 283.54147 3 | 15270 1 | | 76811 146 | 10 902 | | 195.76957 3 | 15908 5 | |
| 3·5 | 41178 431 | 9 830 | | 294.22135 0 | 15070 0 | | 77353 964 | − 10 462 | | + 206.35870 2 | 15652 5 | |
| 3·6 | 40894 760 | 9 269 | | 305.05194 3 | 14876 2 | | 77886 316 | 10 046 | | 217.10438 0 | 15409 5 | |
| 3·7 | 40620 366 | 8 754 | | 316.03130 9 | 14689 5 | | 78408 618 | 9 654 | | 228.00417 4 | 15177 9 | |
| 3·8 | 40354 733 | 8 276 | | 327.15758 1 | 14508 8 | | 78921 263 | 9 279 | | 239.05576 7 | 14957 5 | |
| 3·9 | 40097 383 | 7 835 | | 338.42895 2 | 14334 1 | | 79424 625 | 8 927 | | 250.25695 3 | 14746 8 | |
| 4·0 | 39847 874 | 7 428 | | ★349.84367 5 | 14165 4 | | 79919 057 | − 8 591 | | + 261.60562 4 | 14545 5 | |
| 4·1 | 39605 798 | 7 045 | | 1.40006 1 | 14001 5 | | 80404 895 | 8 274 | | 273.09976 5 | 14352 4 | |
| 4·2 | 39370 773 | 6 695 | | 13.09647 2 | 13843 5 | | 80882 456 | 7 972 | | 284.73744 5 | 14167 7 | |
| 4·3 | 39142 447 | 6 365 | | 24.93132 6 | 13689 7 | | 81352 042 | 7 686 | | 296.51681 5 | 13990 0 | |
| 4·4 | 38920 490 | 6 060 | | 36.90308 6 | 13541 1 | | 81813 939 | 7 416 | | 308.43609 8 | 13819 6 | |
| 4·5 | 38704 596 | 5 772 | | 49.01026 5 | 13396 7 | | 82268 418 | − 7 156 | | + 320.49358 8 | 13655 0 | |
| 4·6 | 38494 478 | 5 506 | | 61.25141 9 | 13256 8 | | 82715 738 | 6 913 | | 332.68764 0 | 13497 2 | |
| 4·7 | 38289 869 | 5 256 | | 73.62514 8 | 13120 7 | | 83156 143 | 6 680 | | 345.01667 4 | 13344 4 | |
| 4·8 | 38090 519 | 5 021 | | 86.13009 1 | 12988 6 | | 83589 866 | 6 457 | | ★357.47916 2 | 13197 4 | |
| 4·9 | 37896 193 | 4 802 | | 98.76492 7 | 12860 5 | | 84017 130 | 6 247 | | 10.07363 3 | 13055 1 | |
| 5·0 | 37706 671 | 4 595 | | 111.52837 4 | 12735 7 | | 84438 145 | − 6 048 | | + 22.79866 4 | 12917 6 | |
| | | | | 1 rev. + | | | | | | 1 rev. + | | |

Interpolation formulae:
$$\begin{cases} \text{When } |\gamma_1^4 - \gamma_0^4| > 1 & \text{use} \quad f_\theta = (1-\theta)f_0 + \theta f_1 + E_0^2 \delta_{m0}^2 + E_1^2 \delta_{m1}^2 + M_0^4 \gamma_0^4 + M_1^4 \gamma_1^4 \\ \text{When } |\gamma_1^4 - \gamma_0^4| \leqslant 1 & \text{use} \quad f_\theta = (1-\theta)f_0 + \theta f_1 + E_0^2 \delta_{m0}^2 + E_1^2 \delta_{m1}^2 + T^4(\gamma_0^4 + \gamma_1^4) \end{cases}$$

TABLE VII—AUXILIARY FUNCTIONS

| $x$ | $F(-x)$ | $\delta^2$ | $\chi(-x)$ | $\delta^2_m$ | $G(-x)$ | $\delta^2$ | $\psi(-x)$ | $\delta^2_m$ |
|---|---|---|---|---|---|---|---|---|
| | | | 1 rev. + | | | | 1 rev. + | |
| 5·0 | 0·37706 671 | +4 597 | 111·52837 4 | +12735 7 | 0·84438 145 | −6 049 | 22·79866 4 | +12917 6 |
| 5·1 | ·37521 746 | 4 403 | 124·41918 4 | 12614 2 | ·84853 111 | 5 855 | 35·65287 9 | 12784 7 |
| 5·2 | ·37341 224 | 4 221 | 137·43614 2 | 12496 1 | ·85262 222 | 5 675 | 48·63494 8 | 12655 6 |
| 5·3 | ·37164 923 | 4 047 | 150·57806 7 | 12381 4 | ·85665 658 | 5 499 | 61·74358 1 | 12531 1 |
| 5·4 | ·36992 669 | 3 887 | 163·84381 1 | 12269 2 | ·86063 595 | 5 334 | 74·97753 1 | 12409 7 |
| 5·5 | 0·36824 302 | +3 732 | 177·23225 3 | +12160 2 | 0·86456 198 | −5 174 | 88·33558 5 | +12292 3 |
| 5·6 | ·36659 667 | 3 587 | 190·74230 2 | 12054 1 | ·86843 627 | 5 022 | 101·81656 8 | 12178 2 |
| 5·7 | ·36498 619 | 3 452 | 204·37289 6 | 11950 2 | ·87226 034 | 4 877 | 115·41933 9 | 12067 2 |
| 5·8 | ·36341 023 | 3 321 | 218·12299 7 | 11849 3 | ·87603 564 | 4 739 | 129·14278 8 | 11959 6 |
| 5·9 | ·36186 748 | 3 198 | 231·99159 5 | 11750 5 | ·87976 355 | 4 605 | 142·98583 8 | 11854 5 |
| 6·0 | 0·36035 671 | +3 083 | 245·97770 3 | +11654 4 | 0·88344 541 | −4 477 | 156·94743 9 | +11752 6 |
| 6·1 | ·35887 677 | 2 971 | 260·08035 9 | 11560 1 | ·88708 250 | 4 356 | 171·02657 1 | 11653 3 |
| 6·2 | ·35742 654 | 2 867 | 274·29862 1 | 11468 6 | ·89067 603 | 4 238 | 185·22224 0 | 11556 2 |
| 6·3 | ·35600 498 | 2 767 | 288·63157 2 | 11378 7 | ·89422 718 | 4 126 | 199·53347 6 | 11461 8 |
| 6·4 | ·35461 109 | 2 673 | 303·07831 4 | 11291 1 | ·89773 707 | 4 017 | 213·95933 5 | 11370 1 |
| 6·5 | 0·35324 393 | +2 582 | 317·63797 0 | +11205 2 | 0·90120 679 | −3 915 | 228·49889 8 | +11280 0 |
| 6·6 | ·35190 259 | 2 496 | 332·30968 2 | 11121 6 | ·90463 736 | 3 814 | 243·15126 5 | 11192 5 |
| 6·7 | ·35058 621 | 2 415 | *347·09261 3 | 11039 2 | ·90802 979 | 3 719 | 257·91556 0 | 11106 7 |
| 6·8 | ·34929 398 | 2 336 | 1·98594 0 | 10959 2 | ·91138 503 | 3 626 | 272·79092 6 | 11023 2 |
| 6·9 | ·34802 511 | 2 262 | 16·98886 2 | 10880 6 | ·91470 401 | 3 537 | 287·77652 7 | 10941 5 |
| 7·0 | 0·34677 886 | +2 192 | 32·10059 3 | +10803 5 | 0·91798 762 | −3 453 | 302·87154 6 | +10861 6 |
| 7·1 | ·34555 453 | 2 121 | 47·32036 2 | 10728 2 | ·92123 670 | 3 370 | 318·07518 4 | 10783 4 |
| 7·2 | ·34435 141 | 2 059 | 62·64741 6 | 10654 3 | ·92445 208 | 3 290 | 333·38665 9 | 10707 0 |
| 7·3 | ·34316 888 | 1 996 | 78·08101 6 | 10582 2 | ·92763 456 | 3 215 | *348·80520 7 | 10632 1 |
| 7·4 | ·34200 631 | 1 935 | 93·62044 0 | 10510 9 | ·93078 489 | 3 140 | 4·33007 9 | 10559 1 |
| 7·5 | 0·34086 309 | +1 880 | 109·26497 6 | +10441 7 | 0·93390 382 | −3 069 | 19·96054 4 | +10487 2 |
| 7·6 | ·33973 867 | 1 824 | 125·01393 1 | 10373 3 | ·93699 206 | 3 001 | 35·69588 4 | 10417 0 |
| 7·7 | ·33863 249 | 1 772 | 140·86662 2 | 10306 5 | ·94005 029 | 2 934 | 51·53539 7 | 10348 3 |
| 7·8 | ·33754 403 | 1 723 | 156·82230 0 | 10240 9 | ·94307 918 | 2 869 | 67·47839 5 | 10280 6 |
| 7·9 | ·33647 280 | 1 672 | 172·88054 9 | 10176 4 | ·94607 938 | 2 809 | 83·52420 2 | 10214 7 |
| 8·0 | 0·33541 829 | +1 629 | 189·04048 4 | +10113 2 | 0·94905 149 | −2 748 | 99·67215 8 | +10149 8 |
| 8·1 | ·33438 007 | 1 582 | 205·30155 3 | 10051 2 | ·95199 612 | 2 690 | 115·92161 4 | 10086 2 |
| 8·2 | ·33335 767 | 1 540 | 221·66313 6 | 9990 2 | ·95491 385 | 2 633 | 132·27193 4 | 10023 8 |
| 8·3 | ·33235 067 | 1 500 | 238·12462 3 | 9930 1 | ·95780 525 | 2 581 | 148·72249 4 | 9962 5 |
| 8·4 | ·33135 867 | 1 458 | 254·68541 4 | 9871 7 | ·96067 084 | 2 526 | 165·27268 1 | 9902 3 |
| 8·5 | 0·33038 125 | +1 422 | 271·34492 3 | + 9813 6 | 0·96351 117 | −2 477 | 181·92189 3 | + 9843 4 |
| 8·6 | ·32941 805 | 1 385 | 288·10257 0 | 9757 0 | ·96632 673 | 2 426 | 198·66954 1 | 9785 4 |
| 8·7 | ·32846 870 | 1 350 | 304·95778 8 | 9701 0 | ·96911 803 | 2 380 | 215·51504 5 | 9728 3 |
| 8·8 | ·32753 285 | 1 315 | 321·91001 8 | 9646 3 | ·97188 553 | 2 331 | 232·45783 4 | 9672 4 |
| 8·9 | ·32661 015 | 1 282 | 338·95871 2 | 9591 9 | ·97462 972 | 2 289 | 249·49734 9 | 9617 6 |
| 9·0 | 0·32570 027 | +1 251 | *356·10332 8 | + 9539 3 | 0·97735 102 | −2 243 | 266·63304 1 | + 9563 3 |
| 9·1 | ·32480 290 | 1 221 | 13·34333 8 | 9486 7 | ·98004 989 | 2 201 | 283·86436 8 | 9510 4 |
| 9·2 | ·32391 774 | 1 190 | 30·67821 7 | 9435 5 | ·98272 675 | 2 161 | 301·19080 0 | 9457 9 |
| 9·3 | ·32304 448 | 1 163 | 48·10745 2 | 9385 0 | ·98538 200 | 2 119 | 318·61181 3 | 9406 5 |
| 9·4 | ·32218 285 | 1 135 | 65·63053 8 | 9335 2 | ·98801 606 | 2 083 | 336·12689 3 | 9356 1 |
| 9·5 | 0·32133 257 | +1 109 | 83·24697 7 | + 9286 0 | 0·99062 929 | −2 043 | *353·73553 5 | + 9306 4 |
| 9·6 | ·32049 338 | 1 082 | 100·95627 8 | 9237 9 | ·99322 209 | 2 006 | 11·43724 2 | 9257 2 |
| 9·7 | ·31966 501 | 1 058 | 118·75795 9 | 9190 5 | ·99579 483 | 1 973 | 29·23152 3 | 9209 2 |
| 9·8 | ·31884 722 | 1 033 | 136·65154 6 | 9143 7 | 0·99834 784 | 1 936 | 47·11789 7 | 9161 6 |
| 9·9 | ·31803 976 | 1 012 | 154·63657 1 | 9097 4 | 1·00088 149 | 1 902 | 65·09588 9 | 9115 1 |
| 10·0 | 0·31724 242 | + 988 | 172·71257 2 | + 9052 2 | 1·00339 612 | −1 871 | 83·16503 3 | + 9068 9 |
| | | | 3 rev. + | | | | 3 rev. + | |

# TABLE VII—AUXILIARY FUNCTIONS

| $x$ | $F(-x)$ | $\delta^2$ | $\chi(-x)$ | $\delta^2$ | $G(-x)$ | $\delta^2$ | $\psi(-x)$ | $\delta^2$ |
|---|---|---|---|---|---|---|---|---|
| | | | 3 rev. + | | | | 3 rev. + | |
| | | | ° | | | | ° | |
| 10·0 | 0·31724 242 | +988 | 172·71257 2 | +9052 3 | 1·00339 612 | − 1 871 | 83·16503 3 | +9069 1 |
| 10·1 | ·31645 496 | 965 | 190·87909 6 | 9007 6 | ·00589 204 | 1 838 | 101·32486 8 | 9024 0 |
| 10·2 | ·31567 715 | 947 | 209·13569 6 | 8963 4 | ·00836 958 | 1 807 | 119·57494 3 | 8979 1 |
| 10·3 | ·31490 881 | 924 | 227·48193 0 | 8920 2 | ·01082 905 | 1 776 | 137·91480 9 | 8935 3 |
| 10·4 | ·31414 971 | 905 | 245·91736 6 | 8877 2 | ·01327 076 | 1 748 | 156·34402 8 | 8892 0 |
| 10·5 | 0·31339 966 | +886 | 264·44157 4 | +8835 0 | 1·01569 499 | − 1 718 | 174·86216 7 | +8849 3 |
| 10·6 | ·31265 847 | 867 | 283·05413 2 | 8793 5 | ·01810 204 | 1 690 | 193·46879 9 | 8807 2 |
| 10·7 | ·31192 595 | 849 | 301·75462 5 | 8752 4 | ·02049 219 | 1 664 | 212·16350 3 | 8765 7 |
| 10·8 | ·31120 192 | 832 | 320·54264 2 | 8711 9 | ·02286 570 | 1 636 | 230·94586 4 | 8724 9 |
| 10·9 | ·31048 621 | 814 | 339·41777 8 | 8672 1 | ·02522 285 | 1 610 | 249·81547 4 | 8684 5 |
| 11·0 | 0·30977 864 | +798 | 358·37963 5 | +8632 7 | 1·02756 390 | − 1 585 | 268·77192 9 | +8644 8 |
| 11·1 | ·30907 905 | 783 | ★ 17·42781 9 | 8593 7 | ·02988 910 | 1 561 | 287·81483 2 | 8605 5 |
| 11·2 | ·30838 729 | 766 | 36·56194 0 | 8555 6 | ·03219 869 | 1 536 | 306·94379 0 | 8566 9 |
| 11·3 | ·30770 319 | 751 | 55·78161 7 | 8517 7 | ·03449 292 | 1 512 | 326·15841 7 | 8528 7 |
| 11·4 | ·30702 660 | 737 | 75·08647 1 | 8480 3 | ·03677 203 | 1 491 | 345·45833 1 | 8491 1 |
| 11·5 | 0·30635 738 | +722 | 94·47612 8 | +8443 6 | 1·03903 623 | − 1 466 | ★ 4·84315 6 | +8453 8 |
| 11·6 | ·30569 538 | 709 | 113·95022 1 | 8407 1 | ·04128 577 | 1 446 | 24·31251 9 | 8417 3 |
| 11·7 | ·30504 047 | 695 | 133·50838 5 | 8371 2 | ·04352 085 | 1 424 | 43·86605 5 | 8381 0 |
| 11·8 | ·30439 251 | 681 | 153·15026 1 | 8335 9 | ·04574 169 | 1 403 | 63·50340 1 | 8345 2 |
| 11·9 | ·30375 136 | 669 | 172·87549 6 | 8300 8 | ·04794 850 | 1 383 | 83·22419 9 | 8310 1 |
| 12·0 | 0·30311 690 | +657 | 192·68373 9 | +8266 2 | 1·05014 148 | − 1 362 | 103·02809 8 | +8275 2 |
| 12·1 | ·30248 901 | 645 | 212·57464 4 | 8232 2 | ·05232 084 | 1 343 | 122·91474 9 | 8240 7 |
| 12·2 | ·30186 757 | 632 | 232·54787 1 | 8198 3 | ·05448 677 | 1 324 | 142·88380 7 | 8206 9 |
| 12·3 | ·30125 245 | 621 | 252·60308 1 | 8165 1 | ·05663 946 | 1 306 | 162·93493 4 | 8173 2 |
| 12·4 | ·30064 354 | 610 | 272·73994 2 | 8132 2 | ·05877 909 | 1 286 | 183·06779 3 | 8140 2 |
| 12·5 | 0·30004 073 | +600 | 292·95812 5 | +8099 6 | 1·06090 586 | − 1 270 | 203·28205 4 | +8107 4 |
| 12·6 | ·29944 392 | 587 | 313·25730 4 | 8067 5 | ·06301 993 | 1 251 | 223·57738 9 | 8075 0 |
| 12·7 | ·29885 298 | 579 | 333·63715 8 | 8035 8 | ·06512 149 | 1 235 | 243·95347 4 | 8043 2 |
| 12·8 | ·29826 783 | 568 | ★ 354·09737 0 | 8004 4 | ·06721 070 | 1 217 | 264·40999 1 | 8011 5 |
| 12·9 | ·29768 836 | 558 | 14·63762 6 | 7973 4 | ·06928 774 | 1 202 | 284·94662 3 | 7980 2 |
| 13·0 | 0·29711 447 | +549 | 35·25761 6 | +7942 7 | 1·07135 276 | − 1 186 | 305·56305 7 | +7949 6 |
| 13·1 | ·29654 607 | 539 | 55·95703 3 | 7912 3 | ·07340 592 | 1 169 | 326·25898 7 | 7918 9 |
| 13·2 | ·29598 306 | 530 | 76·73557 3 | 7882 5 | ·07544 739 | 1 154 | ★ 347·03410 6 | 7888 9 |
| 13·3 | ·29542 535 | 521 | 97·59293 8 | 7852 9 | ·07747 732 | 1 139 | 7·88811 4 | 7859 0 |
| 13·4 | ·29487 285 | 512 | 118·52883 2 | 7823 4 | ·07949 586 | 1 125 | 28·82071 2 | 7829 6 |
| 13·5 | 0·29432 547 | +505 | 139·54296 0 | +7794 5 | 1·08150 315 | − 1 110 | 49·83160 6 | +7800 4 |
| 13·6 | ·29378 314 | 495 | 160·63503 3 | 7765 9 | ·08349 934 | 1 096 | 70·92050 4 | 7771 7 |
| 13·7 | ·29324 576 | 488 | 181·80476 5 | 7737 5 | ·08548 457 | 1 081 | 92·08711 9 | 7743 1 |
| 13·8 | ·29271 326 | 480 | 203·05187 2 | 7709 5 | ·08745 899 | 1 069 | 113·33116 5 | 7715 0 |
| 13·9 | ·29218 556 | 471 | 224·37607 4 | 7681 8 | ·08942 272 | 1 054 | 134·65236 1 | 7687 2 |
| 14·0 | 0·29166 257 | +466 | 245·77709 4 | +7654 3 | 1·09137 591 | − 1 042 | 156·05042 9 | +7659 4 |
| 14·1 | ·29114 424 | 456 | 267·25465 7 | 7627 2 | ·09331 868 | 1 029 | 177·52509 1 | 7632 4 |
| 14·2 | ·29063 047 | 450 | 288·80849 2 | 7600 3 | ·09525 116 | 1 017 | 199·07607 7 | 7605 2 |
| 14·3 | ·29012 120 | 443 | 310·43833 0 | 7573 7 | ·09717 347 | 1 003 | 220·70311 5 | 7578 6 |
| 14·4 | ·28961 636 | 436 | 332·14390 5 | 7547 5 | ·09908 575 | 992 | 242·40593 9 | 7552 1 |
| 14·5 | 0·28911 588 | +429 | ★ 353·92495 5 | +7521 3 | 1·10098 811 | − 980 | 264·18428 4 | +7526 1 |
| 14·6 | ·28861 969 | 423 | 15·78121 8 | 7495 7 | ·10288 067 | 968 | 286·03789 0 | 7500 1 |
| 14·7 | ·28812 773 | 416 | 37·71243 8 | 7470 2 | ·10476 355 | 957 | 307·96649 7 | 7474 6 |
| 14·8 | ·28763 993 | 411 | 59·71836 0 | 7444 8 | ·10663 686 | 946 | 329·96985 0 | 7449 1 |
| 14·9 | ·28715 624 | 402 | 81·79873 0 | 7420 0 | ·10850 071 | 934 | 352·04769 4 | 7424 2 |
| 15·0 | 0·28667 657 | +399 | 103·95330 0 | +7395 1 | 1·11035 522 | − 923 | ★ 14·19978 0 | +7399 2 |
| | | | 6 rev. + | | | | 6 rev. + | |

TABLE VII—AUXILIARY FUNCTIONS

| $x$ | $F(-x)$ | $\delta^2$ | $\chi(-x)$ | $\delta^2$ | $G(-x)$ | $\delta^2$ | $\psi(-x)$ | $\delta^2$ |
|---|---|---|---|---|---|---|---|---|
| | | | 6 rev. + | | | | 6 rev. + | |
| 15·0 | 0·28667 657 | +399 | 103·95330 0 | +7395 1 | 1·11035 522 | −923 | 14·19978 0 | +7399 2 |
| 15·1 | ·28620 089 | 392 | 126·18182 1 | 7370 6 | ·11220 050 | 914 | 36·42585 8 | 7374 7 |
| 15·2 | ·28572 913 | 385 | 148·48404 8 | 7346 5 | ·11403 664 | 902 | 58·72568 3 | 7350 4 |
| 15·3 | ·28526 122 | 382 | 170·85974 0 | 7322 4 | ·11586 376 | 893 | 81·09901 2 | 7326 1 |
| 15·4 | ·28479 713 | 374 | 193·30865 6 | 7298 6 | ·11768 195 | 881 | 103·54560 2 | 7302 4 |
| 15·5 | 0·28433 678 | +369 | 215·83055 8 | +7275 1 | 1·11949 133 | −873 | 126·06521 6 | +7278 8 |
| 15·6 | ·28388 012 | 365 | 238·42521 1 | 7251 8 | ·12129 198 | 863 | 148·65761 8 | 7255 2 |
| 15·7 | ·28342 711 | 359 | 261·09238 2 | 7228 5 | ·12308 400 | 852 | 171·32257 2 | 7232 2 |
| 15·8 | ·28297 769 | 354 | 283·83183 8 | 7205 8 | ·12486 750 | 844 | 194·05984 8 | 7209 2 |
| 15·9 | ·28253 181 | 349 | 306·64335 2 | 7183 1 | ·12664 256 | 834 | 216·86921 6 | 7186 4 |
| 16·0 | 0·28208 942 | +343 | 329·52669 7 | +7160 6 | 1·12840 928 | −826 | 239·75044 8 | +7163 9 |
| 16·1 | ·28165 046 | 341 | ₓ352·48164 8 | 7138 4 | ·13016 774 | 816 | 262·70331 9 | 7141 6 |
| 16·2 | ·28121 491 | 333 | *15·50798 3 | 7116 4 | ·13191 804 | 807 | 285·72760 6 | 7119 4 |
| 16·3 | ·28078 269 | 331 | 38·60548 2 | 7094 4 | ·13366 027 | 800 | 308·82308 7 | 7097 6 |
| 16·4 | ·28035 378 | 325 | 61·77392 5 | 7072 9 | ·13539 450 | 790 | 331·98954 4 | 7075 9 |
| 16·5 | 0·27992 812 | +321 | 85·01309 7 | +7051 4 | 1·13712 083 | −782 | ₓ355·22676 0 | +7054 3 |
| 16·6 | ·27950 567 | 318 | 108·32278 3 | 7030 2 | ·13883 934 | 773 | *18·53451 9 | 7033 1 |
| 16·7 | ·27908 640 | 311 | 131·70277 1 | 7009 1 | ·14055 012 | 767 | 41·91260 9 | 7011 9 |
| 16·8 | ·27867 024 | 309 | 155·15285 0 | 6988 2 | ·14225 323 | 757 | 65·36081 8 | 6991 0 |
| 16·9 | ·27825 717 | 304 | 178·67281 1 | 6967 5 | ·14394 877 | 751 | 88·87893 7 | 6970 2 |
| 17·0 | 0·27784 714 | +300 | 202·26244 7 | +6947 1 | 1·14563 680 | −742 | 112·46675 8 | +6949 7 |
| 17·1 | ·27744 011 | 297 | 225·92155 4 | 6926 8 | ·14731 741 | 734 | 136·12407 6 | 6929 4 |
| 17·2 | ·27703 605 | 292 | 249·64992 9 | 6906 5 | ·14899 068 | 728 | 159·85068 8 | 6909 0 |
| 17·3 | ·27663 491 | 289 | 273·44736 9 | 6886 6 | ·15065 667 | 720 | 183·64639 0 | 6889 2 |
| 17·4 | ·27623 666 | 285 | 297·31367 5 | 6866 8 | ·15231 546 | 713 | 207·51098 4 | 6869 2 |
| 17·5 | 0·27584 126 | +281 | 321·24864 9 | +6847 2 | 1·15396 712 | −706 | 231·44427 0 | +6849 6 |
| 17·6 | ·27544 867 | 278 | ₓ345·25209 5 | 6827 7 | ·15561 172 | 698 | 255·44605 2 | 6830 0 |
| 17·7 | ·27505 886 | 274 | *9·32381 8 | 6808 4 | ·15724 934 | 692 | 279·51613 4 | 6810 7 |
| 17·8 | ·27467 179 | 270 | 33·46362 5 | 6789 3 | ·15888 004 | 685 | 303·65432 3 | 6791 6 |
| 17·9 | ·27428 742 | 268 | 57·67132 5 | 6770 2 | ·16050 389 | 679 | 327·86042 8 | 6772 4 |
| 18·0 | 0·27390 573 | +264 | 81·94672 7 | +6751 6 | 1·16212 095 | −671 | ₓ352·13425 7 | +6753 8 |
| 18·1 | ·27352 668 | 261 | 106·28964 5 | 6732 8 | ·16373 130 | 666 | *16·47562 4 | 6734 8 |
| 18·2 | ·27315 024 | 258 | 130·69989 1 | 6714 3 | ·16533 499 | 659 | 40·88433 9 | 6716 5 |
| 18·3 | ·27277 638 | 254 | 155·17728 0 | 6695 9 | ·16693 209 | 652 | 65·36021 9 | 6698 0 |
| 18·4 | ·27240 506 | 251 | 179·72162 8 | 6677 8 | ·16852 267 | 647 | 89·90307 9 | 6679 7 |
| 18·5 | 0·27203 625 | +248 | 204·33275 4 | +6659 7 | 1·17010 678 | −641 | 114·51273 6 | +6661 7 |
| 18·6 | ·27166 992 | 246 | 229·01047 7 | 6641 8 | ·17168 448 | 634 | 139·18901 0 | 6643 7 |
| 18·7 | ·27130 605 | 242 | 253·75461 8 | 6623 9 | ·17325 584 | 628 | 163·93172 1 | 6625 9 |
| 18·8 | ·27094 460 | 239 | 278·56499 8 | 6606 5 | ·17482 092 | 623 | 188·74069 1 | 6608 2 |
| 18·9 | ·27058 554 | 237 | 303·44144 3 | 6588 8 | ·17637 977 | 617 | 213·61574 3 | 6590 8 |
| 19·0 | 0·27022 885 | +234 | 328·38377 6 | +6571 6 | 1·17793 245 | −611 | 238·55670 3 | +6573 3 |
| 19·1 | ·26987 450 | 231 | ₓ353·39182 5 | 6554 3 | ·17947 902 | 606 | 263·56339 6 | 6556 1 |
| 19·2 | ·26952 246 | 229 | *18·46541 7 | 6537 3 | ·18101 953 | 600 | 288·63565 0 | 6538 9 |
| 19·3 | ·26917 271 | 225 | 43·60438 2 | 6520 2 | ·18255 404 | 595 | 313·77329 3 | 6522 1 |
| 19·4 | ·26882 521 | 223 | 68·80854 9 | 6503 6 | ·18408 260 | 589 | 338·97615 7 | 6505 1 |
| 19·5 | 0·26847 994 | +220 | 94·07775 2 | +6486 7 | 1·18560 527 | −585 | *4·24407 2 | +6488 5 |
| 19·6 | ·26813 687 | 219 | 119·41182 2 | 6470 3 | ·18712 209 | 578 | 29·57687 2 | 6471 8 |
| 19·7 | ·26779 599 | 215 | 144·81059 5 | 6453 8 | ·18863 313 | 574 | 54·97439 0 | 6455 3 |
| 19·8 | ·26745 726 | 214 | 170·27390 6 | 6437 4 | ·19013 843 | 569 | 80·43646 1 | 6439 2 |
| 19·9 | ·26712 067 | 210 | 195·80159 1 | 6421 4 | ·19163 804 | 564 | 105·96292 4 | 6422 8 |
| 20·0 | 0·26678 618 | +208 | 221·39349 0 | +6405 3 | 1·19313 201 | −558 | 131·55361 5 | +6406 7 |
| | | | 9 rev. + | | | | 9 rev. + | |

Table VII—Auxiliary Functions

| $x$ | $F(-x)$ | $\delta^2$ | $\chi(-x)$ | $\delta^2$ | $G(-x)$ | $\delta^2$ | $\psi(-x)$ | $\delta^2$ |
|---|---|---|---|---|---|---|---|---|
| | | | 9 rev. + | | | | 9 rev. + | |
| 20·0 | 0·26678 618 | +208 | 221·39349 0 | +6405 3 | 1·19313 201 | −558 | 131·55361 5 | +6406 7 |
| 20·1 | ·26645 377 | 207 | 247·04944 2 | 6389 2 | ·19462 040 | 555 | 157·20837 3 | 6390 8 |
| 20·2 | ·26612 343 | 203 | 272·76928 6 | 6373 5 | ·19610 324 | 549 | 182·92703 9 | 6374 9 |
| 20·3 | ·26579 512 | 201 | 298·55286 5 | 6357 8 | ·19758 059 | 544 | 208·70945 4 | 6359 2 |
| 20·4 | ·26546 882 | 200 | 324·40002 2 | 6342 2 | ·19905 250 | 540 | 234·55546 1 | 6343 6 |
| 20·5 | 0·26514 452 | +197 | 350·31060 1 | +6326 6 | 1·20051 901 | −536 | 260·46490 4 | +6328 1 |
| 20·6 | ·26482 219 | 195 | *16·28444 6 | 6311 4 | ·20198 016 | 530 | 286·43762 8 | 6312 6 |
| 20·7 | ·26450 181 | 193 | 42·32140 5 | 6296 1 | ·20343 601 | 526 | 312·47347 8 | 6297 5 |
| 20·8 | ·26418 336 | 191 | 68·42132 5 | 6280 9 | ·20488 660 | 522 | *338·57230 3 | 6282 2 |
| 20·9 | ·26386 682 | 188 | 94·58405 4 | 6265 9 | ·20633 197 | 518 | 4·73395 0 | 6267 2 |
| 21·0 | 0·26355 216 | +187 | 120·80944 2 | +6250 9 | 1·20777 216 | −513 | 30·95826 9 | +6252 2 |
| 21·1 | ·26323 937 | 184 | 147·09733 9 | 6236 2 | ·20920 722 | 509 | 57·24511 0 | 6237 3 |
| 21·2 | ·26292 842 | 184 | 173·44759 8 | 6221 4 | ·21063 719 | 504 | 83·59432 4 | 6222 8 |
| 21·3 | ·26261 931 | 180 | 199·86007 1 | 6206 9 | ·21206 212 | 501 | 110·00576 6 | 6207 9 |
| 21·4 | ·26231 200 | 179 | 226·33461 3 | 6192 2 | ·21348 204 | 497 | 136·47928 7 | 6193 6 |
| 21·5 | 0·26200 648 | +177 | 252·87107 7 | +6177 9 | 1·21489 699 | −492 | 163·01474 4 | +6178 9 |
| 21·6 | ·26170 273 | 176 | 279·46932 0 | 6163 6 | ·21630 702 | 489 | 189·61199 0 | 6164 8 |
| 21·7 | ·26140 074 | 173 | 306·12919 9 | 6149 3 | ·21771 216 | 484 | 216·27088 4 | 6150 5 |
| 21·8 | ·26110 048 | 171 | 332·85057 1 | 6135 3 | ·21911 246 | 481 | 242·99128 3 | 6136 4 |
| 21·9 | ·26080 193 | 171 | 359·63329 6 | 6121 2 | ·22050 795 | 477 | 269·77304 6 | 6122 2 |
| 22·0 | 0·26050 509 | +167 | *26·47723 3 | +6107 4 | 1·22189 867 | −472 | 296·61603 1 | +6108 5 |
| 22·1 | ·26020 992 | 167 | 53·38224 4 | 6093 4 | ·22328 467 | 471 | 323·52010 1 | 6094 5 |
| 22·2 | ·25991 642 | 165 | 80·34818 9 | 6079 8 | ·22466 596 | 465 | *350·48511 6 | 6080 8 |
| 22·3 | ·25962 457 | 163 | 107·37493 2 | 6066 0 | ·22604 260 | 462 | 17·51093 9 | 6067 1 |
| 22·4 | ·25933 435 | 161 | 134·46233 5 | 6052 7 | ·22741 462 | 458 | 44·59743 3 | 6053 6 |
| 22·5 | 0·25904 574 | +160 | 161·61026 5 | +6039 0 | 1·22878 206 | −455 | 71·74446 3 | +6040 1 |
| 22·6 | ·25875 873 | 159 | 188·81858 5 | 6025 8 | ·23014 495 | 452 | 98·95189 4 | 6026 7 |
| 22·7 | ·25847 331 | 156 | 216·08716 3 | 6012 4 | ·23150 332 | 447 | 126·21959 2 | 6013 4 |
| 22·8 | ·25818 945 | 155 | 243·41586 5 | 5999 3 | ·23285 722 | 445 | 153·54742 4 | 6000 2 |
| 22·9 | ·25790 714 | 154 | 270·80456 0 | 5986 1 | ·23420 667 | 441 | 180·93525 8 | 5987 0 |
| 23·0 | 0·25762 637 | +152 | 298·25311 6 | +5973 1 | 1·23555 171 | −438 | 208·38296 2 | +5974 2 |
| 23·1 | ·25734 712 | 151 | 325·76140 3 | 5960 2 | ·23689 237 | 434 | 235·89040 8 | 5961 0 |
| 23·2 | ·25706 938 | 149 | 353·32929 2 | 5947 4 | ·23822 869 | 431 | 263·45746 4 | 5948 2 |
| 23·3 | ·25679 313 | 148 | *20·95665 5 | 5934 5 | ·23956 070 | 428 | 291·08400 2 | 5935 5 |
| 23·4 | ·25651 836 | 146 | 48·64336 3 | 5921 8 | ·24088 843 | 425 | 318·76989 5 | 5922 7 |
| 23·5 | 0·25624 505 | +145 | 76·38928 9 | +5909 4 | 1·24221 191 | −422 | *346·51501 5 | +5910 1 |
| 23·6 | ·25597 319 | 144 | 104·19430 9 | 5896 6 | ·24353 117 | 417 | 14·31923 6 | 5897 6 |
| 23·7 | ·25570 277 | 142 | 132·05829 5 | 5884 4 | ·24484 626 | 417 | 42·18243 3 | 5885 1 |
| 23·8 | ·25543 377 | 141 | 159·98112 5 | 5871 8 | ·24615 718 | 411 | 70·10448 1 | 5872 7 |
| 23·9 | ·25516 618 | 139 | 187·96267 3 | 5859 7 | ·24746 399 | 410 | 98·08525 6 | 5860 5 |
| 24·0 | 0·25489 998 | +139 | 216·00281 8 | +5847 4 | 1·24876 670 | −406 | 126·12463 6 | +5848 1 |
| 24·1 | ·25463 517 | 136 | 244·10143 7 | 5835 2 | ·25006 535 | 404 | 154·22249 7 | 5836 1 |
| 24·2 | ·25437 172 | 136 | 272·25840 8 | 5823 2 | ·25135 996 | 400 | 182·37871 9 | 5823 9 |
| 24·3 | ·25410 963 | 135 | 300·47361 1 | 5811 3 | ·25265 057 | 397 | 210·59318 0 | 5812 1 |
| 24·4 | ·25384 889 | 133 | 328·74692 7 | 5799 2 | ·25393 721 | 396 | 238·86576 2 | 5799 9 |
| 24·5 | 0·25358 948 | +132 | *357·07823 5 | +5787 5 | 1·25521 989 | −391 | 267·19634 3 | +5788 3 |
| 24·6 | ·25333 139 | 131 | 25·46741 8 | 5775 6 | ·25649 866 | 389 | 295·58480 7 | 5776 3 |
| 24·7 | ·25307 461 | 129 | 53·91435 7 | 5764 0 | ·25777 354 | 387 | 324·03103 4 | 5764 8 |
| 24·8 | ·25281 912 | 129 | 82·41893 6 | 5752 4 | ·25904 455 | 383 | *352·53490 9 | 5752 9 |
| 24·9 | ·25256 492 | 127 | 110·98103 9 | 5740 8 | ·26031 173 | 382 | 21·09631 3 | 5741 6 |
| 25·0 | 0·25231 199 | +126 | 139·60055 0 | +5729 2 | 1·26157 509 | −378 | 49·71513 3 | +5730 0 |
| | | | 13 rev. + | | | | 13 rev. + | |

TABLE VII—AUXILIARY FUNCTIONS

| $x$ | $F(-x)$ | $\delta^2$ | $\chi(-x)$ | $\delta^2$ | $G(-x)$ | $\delta^2$ | $\psi(-x)$ | $\delta^2$ |
|---|---|---|---|---|---|---|---|---|
| | | | 13 rev. + | | | | 13 rev. + | |
| 25.0 | 0.25231 199 | +126 | 139.60055 0 | +5729 2 | 1.26157 509 | −378 | 49.71513 3 | +5730 0 |
| 25.1 | .25206 032 | 125 | 168.27735 3 | 5718 0 | .26283 467 | 375 | 78.39125 3 | 5718 5 |
| 25.2 | .25180 990 | 125 | 197.01133 6 | 5706 4 | .26409 050 | 374 | 107.12455 8 | 5707 2 |
| 25.3 | .25156 073 | 121 | 225.80238 3 | 5695 3 | .26534 259 | 370 | 135.91493 5 | 5695 9 |
| 25.4 | .25131 277 | 123 | 254.65038 3 | 5684 0 | .26659 098 | 368 | 164.76227 1 | 5684 7 |
| 25.5 | 0.25106 604 | +120 | 283.55522 3 | +5672 9 | 1.26783 569 | −366 | 193.66645 4 | +5673 4 |
| 25.6 | .25082 051 | 120 | 312.51679 2 | 5661 8 | .26907 674 | 362 | 222.62737 1 | 5662 5 |
| 25.7 | .25057 618 | 118 | *341.53497 9 | 5650 7 | .27031 417 | 361 | 251.64491 3 | 5651 4 |
| 25.8 | .25033 303 | 118 | 10.60967 3 | 5639 8 | .27154 799 | 358 | 280.71896 9 | 5640 4 |
| 25.9 | .25009 106 | 116 | 39.74076 5 | 5628 9 | .27277 823 | 356 | 309.84942 9 | 5629 5 |
| 26.0 | 0.24985 025 | +116 | 68.92814 6 | +5618 1 | 1.27400 491 | −353 | 339.03618 4 | +5618 6 |
| 26.1 | .24961 060 | 115 | 98.17170 8 | 5607 3 | .27522 806 | 350 | *8.27912 5 | 5608 0 |
| 26.2 | .24937 210 | 112 | 127.47134 3 | 5596 5 | .27644 771 | 350 | 37.57814 6 | 5597 1 |
| 26.3 | .24913 472 | 114 | 156.82694 3 | 5586 0 | .27766 386 | 345 | 66.93313 8 | 5586 5 |
| 26.4 | .24889 848 | 111 | 186.23840 3 | 5575 4 | .27887 656 | 345 | 96.34399 5 | 5576 0 |
| 26.5 | 0.24866 335 | +110 | 215.70561 7 | +5564 8 | 1.28008 581 | −341 | 125.81061 2 | +5565 3 |
| 26.6 | .24842 932 | 110 | 245.22847 9 | 5554 4 | .28129 165 | 340 | 155.33288 2 | 5555 0 |
| 26.7 | .24819 639 | 109 | 274.80688 5 | 5543 9 | .28249 409 | 336 | 184.91070 2 | 5544 5 |
| 26.8 | .24796 455 | 108 | 304.44073 0 | 5533 6 | .28369 317 | 336 | 214.54396 7 | 5534 1 |
| 26.9 | .24773 379 | 107 | 334.12991 1 | 5523 3 | .28488 889 | 333 | 244.23257 3 | 5523 8 |
| 27.0 | 0.24750 410 | +106 | *3.87432 5 | +5513 1 | 1.28608 128 | −330 | 273.97641 7 | +5513 7 |
| 27.1 | .24727 547 | 105 | 33.67387 0 | 5502 9 | .28727 037 | 329 | 303.77539 8 | 5503 4 |
| 27.2 | .24704 789 | 105 | 63.52844 4 | 5492 8 | .28845 617 | 326 | *333.62941 3 | 5493 2 |
| 27.3 | .24682 136 | 103 | 93.43794 6 | 5482 7 | .28963 871 | 325 | 3.53836 0 | 5483 3 |
| 27.4 | .24659 586 | 102 | 123.40227 5 | 5472 6 | .29081 800 | 322 | 33.50214 0 | 5473 1 |
| 27.5 | 0.24637 138 | +103 | 153.42133 0 | +5462 8 | 1.29199 407 | −320 | 63.52065 1 | +5463 3 |
| 27.6 | .24614 793 | 100 | 183.49501 3 | 5452 9 | .29316 694 | 318 | 93.59379 5 | 5453 3 |
| 27.7 | .24592 548 | 100 | 213.62322 5 | 5442 9 | .29433 663 | 317 | 123.72147 2 | 5443 5 |
| 27.8 | .24570 403 | 100 | 243.80586 6 | 5433 2 | .29550 315 | 311 | 153.90358 4 | 5433 6 |
| 27.9 | .24548 358 | 98 | 274.04283 9 | 5423 5 | .29666 653 | 312 | 184.14003 2 | 5424 0 |
| 28.0 | 0.24526 411 | + 99 | 304.33404 7 | +5413 7 | 1.29782 679 | −310 | 214.43072 0 | +5414 1 |
| 28.1 | .24504 563 | 96 | *334.67939 2 | 5404 2 | .29898 395 | 309 | 244.77554 9 | 5404 7 |
| 28.2 | .24482 811 | 96 | 5.07877 9 | 5394 4 | .30013 802 | 306 | 275.17442 5 | 5394 9 |
| 28.3 | .24461 155 | 96 | 35.53211 0 | 5385 1 | .30128 903 | 305 | 305.62725 0 | 5385 4 |
| 28.4 | .24439 595 | 94 | 66.03929 2 | 5375 4 | .30243 699 | 302 | 336.13392 9 | 5376 0 |
| 28.5 | 0.24418 129 | + 94 | 96.60022 8 | +5366 1 | 1.30358 193 | −302 | *6.69436 8 | +5366 5 |
| 28.6 | .24396 757 | 94 | 127.21482 5 | 5356 7 | .30472 385 | 298 | 37.30847 2 | 5357 1 |
| 28.7 | .24375 479 | 92 | 157.88298 9 | 5347 3 | .30586 279 | 298 | 67.97614 7 | 5347 7 |
| 28.8 | .24354 293 | 91 | 188.60462 6 | 5338 1 | .30699 875 | 294 | 98.69729 9 | 5338 6 |
| 28.9 | .24333 198 | 92 | 219.37964 4 | 5328 8 | .30813 177 | 295 | 129.47183 7 | 5329 1 |
| 29.0 | 0.24312 195 | + 90 | 250.20795 0 | +5319 6 | 1.30926 184 | −291 | 160.29966 6 | +5320 1 |
| 29.1 | .24291 282 | 90 | 281.08945 2 | 5310 5 | .31038 900 | 291 | 191.18069 6 | 5310 8 |
| 29.2 | .24270 459 | 89 | 312.02405 9 | 5301 3 | .31151 325 | 287 | 222.11483 4 | 5301 8 |
| 29.3 | .24249 725 | 87 | *343.01167 9 | 5292 3 | .31263 463 | 288 | 253.10199 0 | 5292 7 |
| 29.4 | .24229 078 | 89 | 14.05222 2 | 5283 4 | .31375 313 | 284 | 284.14207 3 | 5283 7 |
| 29.5 | 0.24208 520 | + 86 | 45.14559 9 | +5274 3 | 1.31486 879 | −283 | 315.23499 3 | +5274 8 |
| 29.6 | .24188 048 | 87 | 76.29171 9 | 5265 4 | .31598 162 | 282 | *346.38066 1 | 5265 7 |
| 29.7 | .24167 663 | 85 | 107.49049 3 | 5256 6 | .31709 163 | 280 | 17.57898 6 | 5257 0 |
| 29.8 | .24147 363 | 86 | 138.74183 3 | 5247 8 | .31819 884 | 278 | 48.82988 1 | 5248 1 |
| 29.9 | .24127 149 | 83 | 170.04565 1 | 5238 8 | .31930 327 | 277 | 80.13325 7 | 5239 4 |
| 30.0 | 0.24107 018 | + 84 | 201.40185 7 | +5230 3 | 1.32040 493 | −275 | 111.48902 7 | +5230 5 |
| | | | 17 rev. + | | | | 17 rev. + | |

TABLE VII—AUXILIARY FUNCTIONS

| x | F(−x) | $\delta^2_m$ | χ(−x) | $\delta^2_m$ | $\gamma^4$ | G(−x) | $\delta^2_m$ | ψ(−x) | $\delta^2_m$ | $\gamma^4$ |
|---|---|---|---|---|---|---|---|---|---|---|
| | | + | Rev. ° | + | + | | − | Rev. ° | + | + |
| 30 | 0·24107 018 | 8 362 | 17 201·40185 7 | 5·22977 8 | 4 | 1·32040 493 | 27 488 | 17 111·48902 7 | 5·23013 7 | 4 |
| 31 | ·23910 216 | 7 768 | 18 157·82634 2 | 5·14477 3 | 4 | ·33127 287 | 25 956 | 18 67·90932 8 | 5·14509 6 | 4 |
| 32 | ·23721 193 | 7 234 | 19 119·39633 9 | 5·06378 5 | 4 | ·34188 101 | 24 553 | 19 29·47546 5 | 5·06407 8 | 4 |
| 33 | ·23539 413 | 6 750 | 20 86·03080 3 | 4·98650 5 | 3 | ·35224 340 | 23 268 | 19 356·10636 1 | 4·98676 2 | 3 |
| 34 | ·23364 391 | 6 312 | 21 57·65240 3 | 4·91265 3 | 3 | ·36237 292 | 22 085 | 20 327·72465 2 | 4·91288 9 | 3 |
| 35 | 0·23195 688 | 5 914 | 22 34·18724 3 | 4·84199 1 | 3 | 1·37228 142 | 20 993 | 21 304·25641 8 | 4·84220 2 | 3 |
| 36 | ·23032 905 | 5 549 | 23 15·56461 9 | 4·77429 1 | 3 | ·38197 984 | 19 985 | 22 285·63093 2 | 4·77448 3 | 3 |
| 37 | ·22875 677 | 5 221 | 24 1·71679 4 | 4·70935 3 | 3 | ·39147 828 | 19 047 | 23 271·78043 7 | 4·70952 6 | 3 |
| 38 | ·22723 674 | 4 914 | 24 352·57879 6 | 4·64699 0 | 2 | ·40078 612 | 18 183 | 24 262·63994 3 | 4·64715 1 | 2 |
| 39 | ·22576 590 | 4 636 | 25 348·08823 3 | 4·58704 8 | 2 | ·40991 203 | 17 373 | 25 258·14704 4 | 4·58719 1 | 2 |
| 40 | 0·22434 146 | 4 379 | 26 348·18513 3 | 4·52935 9 | 2 | 1·41886 411 | 16 621 | 26 258·24175 2 | 4·52949 2 | 2 |
| 41 | ·22296 085 | 4 143 | 27 352·81178 3 | 4·47379 7 | 2 | ·42764 989 | 15 917 | 27 262·86634 3 | 4·47392 0 | 2 |
| 42 | ·22162 170 | 3 925 | 29 1·91259 7 | 4·42023 1 | 2 | ·43627 641 | 15 261 | 28 271·96522 1 | 4·42034 1 | 2 |
| 43 | ·22032 183 | 3 721 | 30 15·43398 7 | 4·36854 1 | 2 | ·44475 024 | 14 646 | 29 285·48478 6 | 4·36864 4 | 2 |
| 44 | ·21905 920 | 3 535 | 31 33·32424 4 | 4·31862 3 | 2 | ·45307 754 | 14 067 | 30 303·37332 1 | 4·31871 9 | 2 |
| 45 | 0·21783 194 | 3 361 | 32 55·53343 2 | 4·27038 0 | 2 | 1·46126 410 | 13 528 | 31 325·58088 3 | 4·27046 7 | 2 |
| 46 | ·21663 831 | 3 198 | 33 82·01329 1 | 4·22371 6 | 1 | ·46931 533 | 13 015 | 32 352·05920 3 | 4·22379 7 | 1 |
| 47 | ·21547 668 | 3 047 | 34 112·71714 2 | 4·17855 0 | 1 | ·47723 635 | 12 535 | 34 22·76159 6 | 4·17862 6 | 1 |
| 48 | ·21434 554 | 2 906 | 35 147·59980 4 | 4·13480 1 | 1 | ·48503 197 | 12 083 | 35 57·64287 6 | 4·13487 3 | 1 |
| 49 | ·21324 348 | 2 775 | 36 186·61751 5 | 4·09240 1 | 1 | ·49270 672 | 11 653 | 36 96·65927 6 | 4·09246 3 | 1 |
| 50 | 0·21216 918 | 2 650 | 37 229·72786 2 | 4·05127 3 | 1 | 1·50026 489 | 11 251 | 37 139·76837 5 | 4·05133 9 | 1 |
| 51 | ·21112 140 | 2 537 | 38 276·88970 7 | 4·01136 9 | 1 | ·50771 052 | 10 865 | 38 186·92903 6 | 4·01142 1 | 1 |
| 52 | ·21009 900 | 2 426 | 39 328·06313 3 | 3·97261 4 | 1 | ·51504 746 | 10 504 | 39 238·10133 2 | 3·97266 9 | 1 |
| 53 | ·20910 088 | 2 326 | 41 23·20937 6 | 3·93496 4 | 1 | ·52227 933 | 10 159 | 40 293·24649 9 | 3·93501 2 | 1 |
| 54 | ·20812 603 | 2 230 | 42 82·29077 6 | 3·89836 3 | 1 | ·52940 958 | 9 834 | 41 352·32687 2 | 3·89841 2 | 1 |
| 55 | 0·20717 349 | 2 140 | 43 145·27072 4 | 3·86276 6 | 1 | 1·53644 147 | 9 519 | 43 55·30584 1 | 3·86280 8 | 1 |
| 56 | ·20624 236 | 2 054 | 44 212·11361 4 | 3·82812 7 | 1 | ·54337 813 | 9 228 | 44 122·14779 5 | 3·82816 6 | 1 |
| 57 | ·20533 178 | 1 974 | 45 282·78479 9 | 3·79440 2 | 1 | ·55022 249 | 8 943 | 45 192·81808 4 | 3·79444 1 | 1 |
| 58 | ·20444 095 | 1 900 | 46 357·25054 7 | 3·76155 3 | 1 | ·55697 739 | 8 679 | 46 267·28297 5 | 3·76158 8 | 1 |
| 59 | ·20356 912 | 1 826 | 48 75·47800 2 | 3·72954 2 | 1 | ·56364 549 | 8 419 | 47 345·50960 9 | 3·72957 6 | 1 |
| 60 | 0·20271 556 | 1 761 | 49 157·43514 7 | 3·69833 7 | 1 | 1·57022 937 | 8 177 | 49 67·46596 7 | 3·69836 8 | 1 |
| 61 | ·20187 961 | 1 694 | 50 243·09077 0 | 3·66789 9 | 1 | ·57673 146 | 7 946 | 50 153·12083 5 | 3·66793 0 | 1 |
| 62 | ·20106 061 | 1 635 | 51 332·41442 8 | 3·63820 3 | 1 | ·58315 408 | 7 720 | 51 242·44376 9 | 3·63823 0 | 1 |
| 63 | ·20025 796 | 1 576 | 53 65·37641 9 | 3·60921 4 | 1 | ·58949 948 | 7 507 | 52 335·40506 4 | 3·60924 3 | 1 |
| 64 | ·19947 108 | 1 521 | 54 161·94775 0 | 3·58091 1 | 1 | ·59576 979 | 7 305 | 54 71·97572 7 | 3·58093 4 | 1 |
| 65 | 0·19869 942 | 1 470 | 55 262·10011 2 | 3·55325 8 | 1 | 1·60196 704 | 7 109 | 55 172·12744 5 | 3·55328 5 | 1 |
| 66 | ·19794 246 | 1 419 | 57 5·80584 9 | 3·52624 2 | 1 | ·60809 319 | 6 920 | 56 275·83256 4 | 3·52626 3 | 1 |
| 67 | ·19719 970 | 1 373 | 58 113·03793 9 | 3·49982 8 | 1 | ·61415 012 | 6 742 | 58 23·06405 8 | 3·49985 0 | 1 |
| 68 | ·19647 067 | 1 328 | 59 223·76996 5 | 3·47400 3 | 1 | ·62013 962 | 6 569 | 59 133·79551 0 | 3·47402 2 | 1 |
| 69 | ·19575 492 | 1 285 | 60 337·97609 7 | 3·44873 7 | 1 | ·62606 342 | 6 403 | 60 248·00108 8 | 3·44875 8 | 1 |
| 70 | 0·19505 202 | 1 244 | 62 95·63106 6 | 3·42401 8 | 1 | 1·63192 318 | 6 244 | 62 5·65552 4 | 3·42403 5 | 1 |
| 71 | ·19436 156 | 1 205 | 63 216·71014 9 | 3·39982 1 | 1 | ·63772 049 | 6 091 | 63 126·73409 2 | 3·39983 8 | 1 |
| 72 | ·19368 315 | 1 168 | 64 341·18914 6 | 3·37612 9 | | ·64345 688 | 5 943 | 64 251·21259 1 | 3·37614 9 | |
| 73 | ·19301 642 | 1 132 | 66 109·04436 2 | 3·35292 9 | | ·64913 383 | 5 802 | 66 19·06732 8 | 3·35294 3 | |
| 74 | ·19236 101 | 1 098 | 67 240·25259 3 | 3·33019 7 | | ·65475 275 | 5 666 | 67 150·27509 5 | 3·33021 2 | |
| 75 | 0·19171 658 | 1 065 | 69 14·79110 5 | 3·30792 3 | | 1·66031 500 | 5 533 | 68 284·81315 8 | 3·30794 0 | |
| 76 | ·19108 280 | 1 032 | 70 152·63762 1 | 3·28608 9 | | ·66582 191 | 5 407 | 70 62·65924 1 | 3·28610 2 | |
| 77 | ·19045 935 | 1 005 | 71 293·77030 5 | 3·26468 3 | | ·67127 474 | 5 285 | 71 203·79150 5 | 3·26469 6 | |
| 78 | ·18984 595 | 974 | 73 78·16774 8 | 3·24369 0 | | ·67667 471 | 5 166 | 72 348·18854 1 | 3·24370 4 | |
| 79 | ·18924 230 | 948 | 74 225·80895 4 | 3·22309 4 | | ·68202 301 | 5 054 | 74 135·82935 4 | 3·22310 6 | |
| 80 | 0·18864 813 | 921 | 76 16·67332 6 | 3·20288 9 | | 1·68732 077 | 4 942 | 75 286·69334 5 | 3·20290 0 | |

Interpolation formula: $f_\theta = (1-\theta) f_0 + \theta f_1 + E_0^2 \delta^2_{m0} + E_1^2 \delta^2_{m1} + T^4 (\gamma_0^4 + \gamma_1^4)$

# BRITISH ASSOCIATION FOR THE ADVANCEMENT OF SCIENCE
## AUXILIARY TABLES    NUMBER I

# COEFFICIENTS IN THE MODIFIED EVERETT INTERPOLATION FORMULA

| $\theta$ | $E_0^2$ | $\delta^2$ | $E_1^2$ | $\delta^2$ | $M_0^4$ | $M_1^4$ | $T^4$ | $\phi$ | $\theta$ | $E_0^2$ | $\delta^2$ | $E_1^2$ | $\delta^2$ | $M_0^4$ | $M_1^4$ | $T^4$ | $\phi$ |
|---|---|---|---|---|---|---|---|---|---|---|---|---|---|---|---|---|---|
| | −0·0 | + | −0·0 | + | | | | | | −0·0 | + | −0·0 | + | | | | |
| 0·00 | 0000 00 | 10 00 | 0000 00 | 0 | −0·000 | +0·000 | −0·000 | 1·00 | 0·50 | 6250 00 | 5 00 | 6250 00 | 5 00 | +0·219 | +0·219 | +0·219 | 0·50 |
| ·01 | 0328 35 | 9 90 | 0166 65 | 10 | ·108 | ·027 | ·041 | 0·99 | ·51 | 6205 85 | 4 90 | 6289 15 | 5 10 | ·248 | ·188 | ·218 | ·49 |
| ·02 | 0646 80 | 9 80 | 0333 20 | 20 | ·207 | ·053 | ·077 | ·98 | ·52 | 6156 80 | 4 80 | 6323 20 | 5 20 | ·276 | ·157 | ·216 | ·48 |
| ·03 | 0955 45 | 9 70 | 0499 55 | 30 | ·297 | ·080 | ·108 | ·97 | ·53 | 6102 95 | 4 70 | 6352 05 | 5 30 | ·302 | ·124 | ·213 | ·47 |
| ·04 | 1254 40 | 9 60 | 0665 60 | 40 | ·377 | ·106 | ·136 | ·96 | ·54 | 6044 40 | 4 60 | 6375 60 | 5 40 | ·328 | ·091 | ·209 | ·46 |
| 0·05 | 1543 75 | 9 50 | 0831 25 | 50 | −0·450 | +0·132 | −0·159 | 0·95 | 0·55 | 5981 25 | 4 50 | 6393 75 | 5 50 | +0·351 | +0·056 | +0·204 | 0·45 |
| ·06 | 1823 60 | 9 40 | 0996 40 | 60 | ·514 | ·158 | ·178 | ·94 | ·56 | 5913 60 | 4 40 | 6406 40 | 5 60 | ·374 | +0·021 | ·197 | ·44 |
| ·07 | 2094 05 | 9 30 | 1160 95 | 70 | ·571 | ·183 | ·194 | ·93 | ·57 | 5841 55 | 4 30 | 6413 45 | 5 70 | ·395 | −0·016 | ·189 | ·43 |
| ·08 | 2355 20 | 9 20 | 1324 80 | 80 | ·620 | ·208 | ·206 | ·92 | ·58 | 5765 20 | 4 20 | 6414 80 | 5 80 | ·414 | ·053 | ·181 | ·42 |
| ·09 | 2607 15 | 9 10 | 1487 85 | 90 | ·662 | ·232 | ·215 | ·91 | ·59 | 5684 65 | 4 10 | 6410 35 | 5 90 | ·432 | ·090 | ·171 | ·41 |
| 0·10 | 2850 00 | 9 00 | 1650 00 | 1 00 | −0·698 | +0·256 | −0·221 | 0·90 | 0·60 | 5600 00 | 4 00 | 6400 00 | 6 00 | +0·448 | −0·128 | +0·160 | 0·40 |
| ·11 | 3083 85 | 8 90 | 1811 15 | 1 10 | ·728 | ·279 | ·225 | ·89 | ·61 | 5511 35 | 3 90 | 6383 65 | 6 10 | ·463 | ·166 | ·148 | ·39 |
| ·12 | 3308 80 | 8 80 | 1971 20 | 1 20 | ·752 | ·301 | ·225 | ·88 | ·62 | 5418 80 | 3 80 | 6361 20 | 6 20 | ·476 | ·205 | ·135 | ·38 |
| ·13 | 3524 95 | 8 70 | 2130 05 | 1 30 | ·770 | ·323 | ·224 | ·87 | ·63 | 5322 45 | 3 70 | 6332 55 | 6 30 | ·487 | ·243 | ·122 | ·37 |
| ·14 | 3732 40 | 8 60 | 2287 60 | 1 40 | ·783 | ·344 | ·220 | ·86 | ·64 | 5222 40 | 3 60 | 6297 60 | 6 40 | ·497 | ·282 | ·108 | ·36 |
| 0·15 | 3931 25 | 8 50 | 2443 75 | 1 50 | −0·791 | +0·364 | −0·214 | 0·85 | 0·65 | 5118 75 | 3 50 | 6256 25 | 6 50 | +0·505 | −0·321 | +0·092 | 0·35 |
| ·16 | 4121 60 | 8 40 | 2598 40 | 1 60 | ·795 | ·382 | ·206 | ·84 | ·66 | 5011 60 | 3 40 | 6208 40 | 6 60 | ·512 | ·359 | ·077 | ·34 |
| ·17 | 4303 55 | 8 30 | 2751 45 | 1 70 | ·794 | ·400 | ·197 | ·83 | ·67 | 4901 05 | 3 30 | 6153 95 | 6 70 | ·517 | ·397 | ·060 | ·33 |
| ·18 | 4477 20 | 8 20 | 2902 80 | 1 80 | ·789 | ·417 | ·186 | ·82 | ·68 | 4787 20 | 3 20 | 6092 80 | 6 80 | ·521 | ·434 | ·044 | ·32 |
| ·19 | 4642 65 | 8 10 | 3052 35 | 1 90 | ·780 | ·433 | ·173 | ·81 | ·69 | 4670 15 | 3 10 | 6024 85 | 6 90 | ·523 | ·470 | ·026 | ·31 |
| 0·20 | 4800 00 | 8 00 | 3200 00 | 2 00 | −0·768 | +0·448 | −0·160 | 0·80 | 0·70 | 4550 00 | 3 00 | 5950 00 | 7 00 | +0·523 | −0·506 | +0·009 | 0·30 |
| ·21 | 4949 35 | 7 90 | 3345 65 | 2 10 | ·753 | ·462 | ·146 | ·79 | ·71 | 4426 85 | 2 90 | 5868 15 | 7 10 | ·522 | ·540 | −0·009 | ·29 |
| ·22 | 5090 80 | 7 80 | 3489 20 | 2 20 | ·734 | ·474 | ·130 | ·78 | ·72 | 4300 80 | 2 80 | 5779 20 | 7 20 | ·520 | ·573 | ·027 | ·28 |
| ·23 | 5224 45 | 7 70 | 3630 55 | 2 30 | ·713 | ·485 | ·114 | ·77 | ·73 | 4171 95 | 2 70 | 5683 05 | 7 30 | ·515 | ·605 | ·045 | ·27 |
| ·24 | 5350 40 | 7 60 | 3769 60 | 2 40 | ·689 | ·495 | ·097 | ·76 | ·74 | 4040 40 | 2 60 | 5579 60 | 7 40 | ·510 | ·635 | ·063 | ·26 |
| 0·25 | 5468 75 | 7 50 | 3906 25 | 2 50 | −0·663 | +0·503 | −0·080 | 0·75 | 0·75 | 3906 25 | 2 50 | 5468 75 | 7 50 | +0·503 | −0·663 | −0·080 | 0·25 |
| ·26 | 5579 60 | 7 40 | 4040 40 | 2 60 | ·635 | ·510 | ·063 | ·74 | ·76 | 3769 60 | 2 40 | 5350 40 | 7 60 | ·495 | ·689 | ·097 | ·24 |
| ·27 | 5683 05 | 7 30 | 4171 95 | 2 70 | ·605 | ·515 | ·045 | ·73 | ·77 | 3630 55 | 2 30 | 5224 45 | 7 70 | ·485 | ·713 | ·114 | ·23 |
| ·28 | 5779 20 | 7 20 | 4300 80 | 2 80 | ·573 | ·520 | ·027 | ·72 | ·78 | 3489 20 | 2 20 | 5090 80 | 7 80 | ·474 | ·734 | ·130 | ·22 |
| ·29 | 5868 15 | 7 10 | 4426 85 | 2 90 | ·540 | ·522 | −0·009 | ·71 | ·79 | 3345 65 | 2 10 | 4949 35 | 7 90 | ·462 | ·753 | ·146 | ·21 |
| 0·30 | 5950 00 | 7 00 | 4550 00 | 3 00 | −0·506 | +0·523 | +0·009 | 0·70 | 0·80 | 3200 00 | 2 00 | 4800 00 | 8 00 | +0·448 | −0·768 | −0·160 | 0·20 |
| ·31 | 6024 85 | 6 90 | 4670 15 | 3 10 | ·470 | ·523 | ·026 | ·69 | ·81 | 3052 35 | 1 90 | 4642 65 | 8 10 | ·433 | ·780 | ·173 | ·19 |
| ·32 | 6091 80 | 6 80 | 4787 20 | 3 20 | ·434 | ·521 | ·044 | ·68 | ·82 | 2902 80 | 1 80 | 4477 20 | 8 20 | ·417 | ·789 | ·186 | ·18 |
| ·33 | 6153 95 | 6 70 | 4901 05 | 3 30 | ·397 | ·517 | ·060 | ·67 | ·83 | 2751 45 | 1 70 | 4303 55 | 8 30 | ·400 | ·794 | ·197 | ·17 |
| ·34 | 6208 40 | 6 60 | 5011 60 | 3 40 | ·359 | ·512 | ·077 | ·66 | ·84 | 2598 40 | 1 60 | 4121 60 | 8 40 | ·382 | ·795 | ·206 | ·16 |
| 0·35 | 6256 25 | 6 50 | 5118 75 | 3 50 | −0·321 | +0·505 | +0·092 | 0·65 | 0·85 | 2443 75 | 1 50 | 3931 25 | 8 50 | +0·364 | −0·791 | −0·214 | 0·15 |
| ·36 | 6297 60 | 6 40 | 5222 40 | 3 60 | ·282 | ·497 | ·108 | ·64 | ·86 | 2287 60 | 1 40 | 3732 40 | 8 60 | ·344 | ·783 | ·220 | ·14 |
| ·37 | 6332 55 | 6 30 | 5322 45 | 3 70 | ·243 | ·487 | ·122 | ·63 | ·87 | 2130 05 | 1 30 | 3524 95 | 8 70 | ·323 | ·770 | ·224 | ·13 |
| ·38 | 6361 20 | 6 20 | 5418 80 | 3 80 | ·205 | ·476 | ·135 | ·62 | ·88 | 1971 20 | 1 20 | 3308 80 | 8 80 | ·301 | ·752 | ·225 | ·12 |
| ·39 | 6383 65 | 6 10 | 5511 35 | 3 90 | ·166 | ·463 | ·148 | ·61 | ·89 | 1811 15 | 1 10 | 3083 85 | 8 90 | ·279 | ·728 | ·225 | ·11 |
| 0·40 | 6400 00 | 6 00 | 5600 00 | 4 00 | −0·128 | +0·448 | +0·160 | 0·60 | 0·90 | 1650 00 | 1 00 | 2850 00 | 9 00 | +0·256 | −0·698 | −0·221 | 0·10 |
| ·41 | 6410 35 | 5 90 | 5684 65 | 4 10 | ·090 | ·432 | ·171 | ·59 | ·91 | 1487 85 | 90 | 2607 15 | 9 10 | ·232 | ·662 | ·215 | ·09 |
| ·42 | 6414 80 | 5 80 | 5765 20 | 4 20 | ·053 | ·414 | ·181 | ·58 | ·92 | 1324 80 | 80 | 2355 20 | 9 20 | ·208 | ·620 | ·206 | ·08 |
| ·43 | 6413 45 | 5 70 | 5841 55 | 4 30 | −0·016 | ·395 | ·189 | ·57 | ·93 | 1160 95 | 70 | 2094 05 | 9 30 | ·183 | ·571 | ·194 | ·07 |
| ·44 | 6406 40 | 5 60 | 5913 60 | 4 40 | +0·021 | ·374 | ·197 | ·56 | ·94 | 0996 40 | 60 | 1823 60 | 9 40 | ·158 | ·514 | ·178 | ·06 |
| 0·45 | 6393 75 | 5 50 | 5981 25 | 4 50 | +0·056 | +0·351 | +0·204 | 0·55 | 0·95 | 0831 25 | 50 | 1543 75 | 9 50 | +0·132 | −0·450 | −0·159 | 0·05 |
| ·46 | 6375 60 | 5 40 | 6044 40 | 4 60 | ·091 | ·328 | ·209 | ·54 | ·96 | 0665 60 | 40 | 1254 40 | 9 60 | ·106 | ·377 | ·136 | ·04 |
| ·47 | 6352 05 | 5 30 | 6102 95 | 4 70 | ·124 | ·302 | ·213 | ·53 | ·97 | 0499 55 | 30 | 0955 45 | 9 70 | ·080 | ·297 | ·108 | ·03 |
| ·48 | 6323 20 | 5 20 | 6156 80 | 4 80 | ·157 | ·276 | ·216 | ·52 | ·98 | 0333 20 | 20 | 0646 80 | 9 80 | ·053 | ·207 | ·077 | ·02 |
| ·49 | 6289 15 | 5 10 | 6205 85 | 4 90 | ·188 | ·248 | ·218 | ·51 | ·99 | 0166 65 | 10 | 0328 35 | 9 90 | ·027 | ·108 | ·041 | ·01 |
| 0·50 | 6250 00 | 5 00 | 6250 00 | 5 00 | +0·219 | +0·219 | +0·219 | 0·50 | 1·00 | 0000 00 | 0 | 0000 00 | 10 00 | +0·000 | −0·000 | −0·000 | 0·00 |

Formula:  when $|\gamma_1^4 - \gamma_0^4| \leqslant 1$,     $f_\theta = \phi f_0 + \theta f_1 + E_0^2 \delta_{m0}^2 + E_1^2 \delta_{m1}^2 + T^4(\gamma_0^4 + \gamma_1^4)$

when $|\gamma^4 - \gamma_0^4| > 1$,     $f_\theta = \phi f_0 + \theta f_1 + E_0^2 \delta_{m0}^2 + E_1^2 \delta_{m1}^2 + M_0^4 \gamma_0^4 + M_1^4 \gamma_1^4$

where $\delta_m^2 = \delta^2 - 0 \cdot 184\delta^4 + 0 \cdot 038082\delta^6 - 0 \cdot 00830\delta^8 + 0 \cdot 0019\delta^{10} - 0 \cdot 0004\delta^{12} + \ldots$

$1000\gamma^4 = \delta^4 - 0 \cdot 27827\delta^6 + 0 \cdot 0685\delta^8 - 0 \cdot 0164\delta^{10} + 0 \cdot 004\delta^{12} - \ldots$

(See B.A. Mathematical Tables, Part-Volume B, *The Airy Integral*, page B7)

Cambridge: University Press, 1946

# BRITISH ASSOCIATION FOR THE ADVANCEMENT OF SCIENCE
## AUXILIARY TABLES     NUMBER II

# TABLE FOR INTERPOLATION WITH REDUCED DERIVATIVES
### COEFFICIENTS FOR THE FUNCTION

| $\theta$ | $\theta^2$ | $\theta^3$ | $\theta^4$ | $\theta^5$ | $\theta^6$ | $\theta^7$ |
|---|---|---|---|---|---|---|
| 0·00 | 0·0000 | 0·00000 0 | 0·00000 000 | | | 0·00 |
| ·01 | ·0001 | ·00000 1 | ·00000 001 | | | |
| ·02 | ·0004 | ·00000 8 | ·00000 016 | | | |
| ·03 | ·0009 | ·00002 7 | ·00000 081 | | | |
| ·04 | ·0016 | ·00006 4 | ·00000 256 | 0·00000 01 | | |
| 0·05 | 0·0025 | 0·00012 5 | 0·00000 625 | 0·00000 03 | | |
| ·06 | ·0036 | ·00021 6 | ·00001 296 | ·00000 08 | | |
| ·07 | ·0049 | ·00034 3 | ·00002 401 | ·00000 17 | | |
| ·08 | ·0064 | ·00051 2 | ·00004 096 | ·00000 33 | | |
| ·09 | ·0081 | ·00072 9 | ·00006 561 | ·00000 59 | 0·00000 1 | |
| 0·10 | 0·0100 | 0·00100 0 | 0·00010 000 | 0·00001 00 | 0·00000 1 | |
| ·11 | ·0121 | ·00133 1 | ·00014 641 | ·00001 61 | ·00000 2 | |
| ·12 | ·0144 | ·00172 8 | ·00020 736 | ·00002 49 | ·00000 3 | |
| ·13 | ·0169 | ·00219 7 | ·00028 561 | ·00003 71 | ·00000 5 | |
| ·14 | ·0196 | ·00274 4 | ·00038 416 | ·00005 38 | ·00000 8 | |
| 0·15 | 0·0225 | 0·00337 5 | 0·00050 625 | 0·00007 59 | 0·00001 1 | |
| ·16 | ·0256 | ·00409 6 | ·00065 536 | ·00010 49 | ·00001 7 | |
| ·17 | ·0289 | ·00491 3 | ·00083 521 | ·00014 20 | ·00002 4 | |
| ·18 | ·0324 | ·00583 2 | ·00104 976 | ·00018 90 | ·00003 4 | 001 |
| ·19 | ·0361 | ·00685 9 | ·00130 321 | ·00024 76 | ·00004 7 | 001 |
| 0·20 | 0·0400 | 0·00800 0 | 0·00160 000 | 0·00032 00 | 0·00006 4 | 001 |
| ·21 | ·0441 | ·00926 1 | ·00194 481 | ·00040 84 | ·00008 6 | 002 |
| ·22 | ·0484 | ·01064 8 | ·00234 256 | ·00051 54 | ·00011 3 | 002 |
| ·23 | ·0529 | ·01216 7 | ·00279 841 | ·00064 36 | ·00014 8 | 003 |
| ·24 | ·0576 | ·01382 4 | ·00331 776 | ·00079 63 | ·00019 1 | 005 |
| 0·25 | 0·0625 | 0·01562 5 | 0·00390 625 | 0·00097 66 | 0·00024 4 | 006 |
| ·26 | ·0676 | ·01757 6 | ·00456 976 | ·00118 81 | ·00030 9 | 008 |
| ·27 | ·0729 | ·01968 3 | ·00531 441 | ·00143 49 | ·00038 7 | 010 |
| ·28 | ·0784 | ·02195 2 | ·00614 656 | ·00172 10 | ·00048 2 | 013 |
| ·29 | ·0841 | ·02438 9 | ·00707 281 | ·00205 11 | ·00059 5 | 017 |
| 0·30 | 0·0900 | 0·02700 0 | 0·00810 000 | 0·00243 00 | 0·00072 9 | 022 |
| ·31 | ·0961 | ·02979 1 | ·00923 521 | ·00286 29 | ·00088 8 | 028 |
| ·32 | ·1024 | ·03276 8 | ·01048 576 | ·00335 54 | ·00107 4 | 034 |
| ·33 | ·1089 | ·03593 7 | ·01185 921 | ·00391 35 | ·00129 1 | 043 |
| ·34 | ·1156 | ·03930 4 | ·01336 336 | ·00454 35 | ·00154 5 | 053 |
| 0·35 | 0·1225 | 0·04287 5 | 0·01500 625 | 0·00525 22 | 0·00183 8 | 064 |
| ·36 | ·1296 | ·04665 6 | ·01679 616 | ·00604 66 | ·00217 7 | 078 |
| ·37 | ·1369 | ·05065 3 | ·01874 161 | ·00693 44 | ·00256 6 | 095 |
| ·38 | ·1444 | ·05487 2 | ·02085 136 | ·00792 35 | ·00301 1 | 114 |
| ·39 | ·1521 | ·05931 9 | ·02313 441 | ·00902 24 | ·00351 9 | 137 |
| 0·40 | 0·1600 | 0·06400 0 | 0·02560 000 | 0·01024 00 | 0·00409 6 | 164 |
| ·41 | ·1681 | ·06892 1 | ·02825 761 | ·01158 56 | ·00475 0 | 195 |
| ·42 | ·1764 | ·07408 8 | ·03111 696 | ·01306 91 | ·00548 9 | 231 |
| ·43 | ·1849 | ·07950 7 | ·03418 801 | ·01470 08 | ·00632 1 | 272 |
| ·44 | ·1936 | ·08518 4 | ·03748 096 | ·01649 16 | ·00725 6 | 319 |
| 0·45 | 0·2025 | 0·09112 5 | 0·04100 625 | 0·01845 28 | 0·00830 4 | 374 |
| ·46 | ·2116 | ·09733 6 | ·04477 456 | ·02059 63 | ·00947 4 | 436 |
| ·47 | ·2209 | ·10382 3 | ·04879 681 | ·02293 45 | ·01077 9 | 507 |
| ·48 | ·2304 | ·11059 2 | ·05308 416 | ·02548 04 | ·01223 1 | 587 |
| ·49 | ·2401 | ·11764 9 | ·05764 801 | ·02824 75 | ·01384 1 | 678 |
| 0·50 | 0·2500 | 0·12500 0 | 0·06250 000 | 0·03125 00 | 0·01562 5 | 781 |

| $\theta$ | $\theta^2$ | $\theta^3$ | $\theta^4$ | $\theta^5$ | $\theta^6$ | $\theta^7$ |
|---|---|---|---|---|---|---|
| 0·50 | 0·2500 | 0·12500 0 | 0·06250 000 | 0·03125 0 | 0·01563 | 0·0078 |
| ·51 | ·2601 | ·13265 1 | ·06765 201 | ·03450 3 | ·01760 | ·0090 |
| ·52 | ·2704 | ·14060 8 | ·07311 616 | ·03802 0 | ·01977 | ·0103 |
| ·53 | ·2809 | ·14887 7 | ·07890 481 | ·04182 0 | ·02216 | ·0117 |
| ·54 | ·2916 | ·15746 4 | ·08503 056 | ·04591 7 | ·02479 | ·0134 |
| 0·55 | 0·3025 | 0·16637 5 | 0·09150 625 | 0·05032 8 | 0·02768 | 0·0152 |
| ·56 | ·3136 | ·17561 6 | ·09834 496 | ·05507 3 | ·03084 | ·0173 |
| ·57 | ·3249 | ·18519 3 | ·10556 001 | ·06016 9 | ·03430 | ·0195 |
| ·58 | ·3364 | ·19511 2 | ·11316 496 | ·06563 6 | ·03807 | ·0221 |
| ·59 | ·3481 | ·20537 9 | ·12117 361 | ·07149 2 | ·04218 | ·0249 |
| 0·60 | 0·3600 | 0·21600 0 | 0·12960 000 | 0·07776 0 | 0·04666 | 0·0280 |
| ·61 | ·3721 | ·22698 1 | ·13845 841 | ·08446 0 | ·05152 | ·0314 |
| ·62 | ·3844 | ·23832 8 | ·14776 336 | ·09161 3 | ·05680 | ·0352 |
| ·63 | ·3969 | ·25004 7 | ·15752 961 | ·09924 4 | ·06252 | ·0394 |
| ·64 | ·4096 | ·26214 4 | ·16777 216 | ·10737 4 | ·06872 | ·0440 |
| 0·65 | 0·4225 | 0·27462 5 | 0·17850 625 | 0·11602 9 | 0·07542 | 0·0490 |
| ·66 | ·4356 | ·28749 6 | ·18974 736 | ·12523 3 | ·08265 | ·0546 |
| ·67 | ·4489 | ·30076 3 | ·20151 121 | ·13501 3 | ·09046 | ·0606 |
| ·68 | ·4624 | ·31443 2 | ·21381 376 | ·14539 3 | ·09887 | ·0672 |
| ·69 | ·4761 | ·32850 9 | ·22667 121 | ·15640 3 | ·10792 | ·0745 |
| 0·70 | 0·4900 | 0·34300 0 | 0·24010 000 | 0·16807 0 | 0·11765 | 0·0824 |
| ·71 | ·5041 | ·35791 1 | ·25411 681 | ·18042 3 | ·12810 | ·0910 |
| ·72 | ·5184 | ·37324 8 | ·26873 856 | ·19349 2 | ·13931 | ·1003 |
| ·73 | ·5329 | ·38901 7 | ·28398 241 | ·20730 7 | ·15133 | ·1105 |
| ·74 | ·5476 | ·40522 4 | ·29986 576 | ·22190 1 | ·16421 | ·1215 |
| 0·75 | 0·5625 | 0·42187 5 | 0·31640 625 | 0·23730 5 | 0·17798 | 0·1335 |
| ·76 | ·5776 | ·43897 6 | ·33362 176 | ·25355 3 | ·19270 | ·1465 |
| ·77 | ·5929 | ·45653 3 | ·35153 041 | ·27067 8 | ·20842 | ·1605 |
| ·78 | ·6084 | ·47455 2 | ·37015 056 | ·28871 7 | ·22520 | ·1757 |
| ·79 | ·6241 | ·49303 9 | ·38950 081 | ·30770 6 | ·24309 | ·1920 |
| 0·80 | 0·6400 | 0·51200 0 | 0·40960 000 | 0·32768 0 | 0·26214 | 0·2097 |
| ·81 | ·6561 | ·53144 1 | ·43046 721 | ·34867 8 | ·28243 | ·2288 |
| ·82 | ·6724 | ·55136 8 | ·45212 176 | ·37074 0 | ·30401 | ·2493 |
| ·83 | ·6889 | ·57178 7 | ·47458 321 | ·39390 4 | ·32694 | ·2714 |
| ·84 | ·7056 | ·59270 4 | ·49787 136 | ·41821 2 | ·35130 | ·2951 |
| 0·85 | 0·7225 | 0·61412 5 | 0·52200 625 | 0·44370 5 | 0·37715 | 0·3206 |
| ·86 | ·7396 | ·63605 6 | ·54700 816 | ·47042 7 | ·40457 | ·3479 |
| ·87 | ·7569 | ·65850 3 | ·57289 761 | ·49842 1 | ·43363 | ·3773 |
| ·88 | ·7744 | ·68147 2 | ·59969 536 | ·52773 2 | ·46440 | ·4087 |
| ·89 | ·7921 | ·70496 9 | ·62742 241 | ·55840 6 | ·49698 | ·4423 |
| 0·90 | 0·8100 | 0·72900 0 | 0·65610 000 | 0·59049 0 | 0·53144 | 0·4783 |
| ·91 | ·8281 | ·75357 1 | ·68574 961 | ·62403 2 | ·56787 | ·5168 |
| ·92 | ·8464 | ·77868 8 | ·71639 296 | ·65908 2 | ·60636 | ·5578 |
| ·93 | ·8649 | ·80435 7 | ·74805 201 | ·69568 8 | ·64699 | ·6017 |
| ·94 | ·8836 | ·83058 4 | ·78074 896 | ·73390 4 | ·68987 | ·6485 |
| 0·95 | 0·9025 | 0·85737 5 | 0·81450 625 | 0·77378 1 | 0·73509 | 0·6983 |
| ·96 | ·9216 | ·88473 6 | ·84934 656 | ·81537 3 | ·78276 | ·7514 |
| ·97 | ·9409 | ·91267 3 | ·88529 281 | ·85873 4 | ·83297 | ·8080 |
| ·98 | ·9604 | ·94119 2 | ·92236 816 | ·90392 1 | ·88584 | ·8681 |
| ·99 | ·9801 | ·97029 9 | ·96059 601 | ·95099 0 | ·94148 | ·9321 |
| 1·00 | 1·0000 | 1·00000 0 | 1·00000 000 | 1·00000 0 | 1·00000 | 1·0000 |

Formulae: $\quad f(x+\theta h) = f(x) + \theta\tau + \theta^2\tau^2 + \ldots + \theta^n\tau^n + \ldots$

$$f'(x+\theta h) = \tau + 2\theta\tau^2 + 3\theta^2\tau^3 + \ldots + n\theta^{n-1}\tau^n + \ldots$$

where $\tau^n \equiv \tau^n f(x) = \dfrac{h^n}{n!}\dfrac{d^n}{dx^n}f(x)$, in which $h$ is the interval of tabulation in $x$.

(See B.A. Mathematical Tables, Part-Volume B, *The Airy Integral*, page B 7)

Cambridge: University Press, 1946

# BRITISH ASSOCIATION FOR THE ADVANCEMENT OF SCIENCE
## AUXILIARY TABLES    NUMBER II

# TABLE FOR INTERPOLATION WITH REDUCED DERIVATIVES
## COEFFICIENTS FOR THE FIRST DERIVATIVE

| $\theta$ | $2\theta$ | $3\theta^2$ | $4\theta^3$ | $5\theta^4$ | $6\theta^5$ | $7\theta^6$ | $8\theta^7$ | $\theta$ | $2\theta$ | $3\theta^2$ | $4\theta^3$ | $5\theta^4$ | $6\theta^5$ | $7\theta^6$ | $8\theta^7$ |
|---|---|---|---|---|---|---|---|---|---|---|---|---|---|---|---|
| 0·00 | 0·00 | 0·0000 | 0·00000 0 |  |  |  |  | 0·50 | 1·00 | 0·7500 | 0·50000 0 | 0·31250 | 0·1875 | 0·109 | 0·06 |
| ·01 | ·02 | ·0003 | ·00000 4 |  |  |  |  | ·51 | ·02 | ·7803 | ·53060 4 | ·33826 | ·2070 | ·123 | ·07 |
| ·02 | ·04 | ·0012 | ·00003 2 | 0·00000 1 |  |  |  | ·52 | ·04 | ·8112 | ·56243 2 | ·36558 | ·2281 | ·138 | ·08 |
| ·03 | ·06 | ·0027 | ·00010 8 | ·00000 4 |  |  |  | ·53 | ·06 | ·8427 | ·59550 8 | ·39452 | ·2509 | ·155 | ·09 |
| ·04 | ·08 | ·0048 | ·00025 6 | ·00001 3 |  |  |  | ·54 | ·08 | ·8748 | ·62985 6 | ·42515 | ·2755 | ·174 | ·11 |
| 0·05 | 0·10 | 0·0075 | 0·00050 0 | 0·00003 1 |  |  |  | 0·55 | 1·10 | 0·9075 | 0·66550 0 | 0·45753 | 0·3020 | 0·194 | 0·12 |
| ·06 | ·12 | ·0108 | ·00086 4 | ·00006 5 |  |  |  | ·56 | ·12 | ·9408 | ·70246 4 | ·49172 | ·3304 | ·216 | ·14 |
| ·07 | ·14 | ·0147 | ·00137 2 | ·00012 0 | 0·00001 |  |  | ·57 | ·14 | 0·9747 | ·74077 2 | ·52780 | ·3610 | ·240 | ·16 |
| ·08 | ·16 | ·0192 | ·00204 8 | ·00020 5 | ·00002 |  |  | ·58 | ·16 | 1·0092 | ·78044 8 | ·56582 | ·3938 | ·266 | ·18 |
| ·09 | ·18 | ·0243 | ·00291 6 | ·00032 8 | ·00004 |  |  | ·59 | ·18 | ·0443 | ·82151 6 | ·60587 | ·4290 | ·295 | ·20 |
| 0·10 | 0·20 | 0·0300 | 0·00400 0 | 0·00050 0 | 0·00006 |  |  | 0·60 | 1·20 | 1·0800 | 0·86400 0 | 0·64800 | 0·4666 | 0·327 | 0·22 |
| ·11 | ·22 | ·0363 | ·00532 4 | ·00073 2 | ·00010 |  |  | ·61 | ·22 | ·1163 | ·90792 4 | ·69229 | ·5068 | ·361 | ·25 |
| ·12 | ·24 | ·0432 | ·00691 2 | ·00103 7 | ·00015 |  |  | ·62 | ·24 | ·1532 | 0·95331 2 | ·73882 | ·5497 | ·398 | ·28 |
| ·13 | ·26 | ·0507 | ·00878 8 | ·00142 8 | ·00022 |  |  | ·63 | ·26 | ·1907 | 1·00018 8 | ·78765 | ·5955 | ·438 | ·32 |
| ·14 | ·28 | ·0588 | ·01097 6 | ·00192 1 | ·00032 | 0·0001 |  | ·64 | ·28 | ·2288 | ·04857 6 | ·83886 | ·6442 | ·481 | ·35 |
| 0·15 | 0·30 | 0·0675 | 0·01350 0 | 0·00253 1 | 0·00046 | 0·0001 |  | 0·65 | 1·30 | 1·2675 | 1·09850 0 | 0·89253 | 0·6962 | 0·528 | 0·39 |
| ·16 | ·32 | ·0768 | ·01638 4 | ·00327 7 | ·00063 | ·0001 |  | ·66 | ·32 | ·3068 | ·14998 4 | 0·94874 | ·7514 | ·579 | ·44 |
| ·17 | ·34 | ·0867 | ·01965 2 | ·00417 6 | ·00085 | ·0002 |  | ·67 | ·34 | ·3467 | ·20305 2 | 1·00756 | ·8101 | ·633 | ·48 |
| ·18 | ·36 | ·0972 | ·02332 8 | ·00524 9 | ·00113 | ·0002 |  | ·68 | ·36 | ·3872 | ·25772 8 | ·06907 | ·8724 | ·692 | ·54 |
| ·19 | ·38 | ·1083 | ·02743 6 | ·00651 6 | ·00149 | ·0003 |  | ·69 | ·38 | ·4283 | ·31403 6 | ·13336 | 0·9384 | ·755 | ·60 |
| 0·20 | 0·40 | 0·1200 | 0·03200 0 | 0·00800 0 | 0·00192 | 0·0004 |  | 0·70 | 1·40 | 1·4700 | 1·37200 0 | 1·20050 | 1·0084 | 0·824 | 0·66 |
| ·21 | ·42 | ·1323 | ·03704 4 | ·00972 4 | ·00245 | ·0006 |  | ·71 | ·42 | ·5123 | ·43164 4 | ·27058 | ·0825 | ·897 | ·73 |
| ·22 | ·44 | ·1452 | ·04259 2 | ·01171 3 | ·00309 | ·0008 |  | ·72 | ·44 | ·5552 | ·49299 2 | ·34369 | ·1610 | 0·975 | ·80 |
| ·23 | ·46 | ·1587 | ·04866 8 | ·01399 2 | ·00386 | ·0010 |  | ·73 | ·46 | ·5987 | ·55606 8 | ·41991 | ·2438 | 1·059 | ·88 |
| ·24 | ·48 | ·1728 | ·05529 6 | ·01658 9 | ·00478 | ·0013 |  | ·74 | ·48 | ·6428 | ·62089 6 | ·49933 | ·3314 | ·149 | 0·97 |
| 0·25 | 0·50 | 0·1875 | 0·06250 0 | 0·01953 1 | 0·00586 | 0·0017 |  | 0·75 | 1·50 | 1·6875 | 1·68750 0 | 1·58203 | 1·4238 | 1·246 | 1·07 |
| ·26 | ·52 | ·2028 | ·07030 4 | ·02284 9 | ·00713 | ·0022 | 0·001 | ·76 | ·52 | ·7328 | ·75590 4 | ·66811 | ·5213 | ·349 | ·17 |
| ·27 | ·54 | ·2187 | ·07873 2 | ·02657 2 | ·00861 | ·0027 | ·001 | ·77 | ·54 | ·7787 | ·82613 2 | ·75765 | ·6241 | ·459 | ·28 |
| ·28 | ·56 | ·2352 | ·08780 8 | ·03073 3 | ·01033 | ·0034 | ·001 | ·78 | ·56 | ·8252 | ·89820 8 | ·85075 | ·7323 | ·576 | ·41 |
| ·29 | ·58 | ·2523 | ·09755 6 | ·03536 4 | ·01231 | ·0042 | ·001 | ·79 | ·58 | ·8723 | 1·97215 6 | 1·94750 | ·8462 | ·702 | ·54 |
| 0·30 | 0·60 | 0·2700 | 0·10800 0 | 0·04050 0 | 0·01458 | 0·0051 | 0·002 | 0·80 | 1·60 | 1·9200 | 2·04800 0 | 2·04800 | 1·9661 | 1·835 | 1·68 |
| ·31 | ·62 | ·2883 | ·11916 4 | ·04617 6 | ·01718 | ·0062 | ·002 | ·81 | ·62 | 1·9683 | ·12576 4 | ·15234 | 2·0921 | 1·977 | 1·83 |
| ·32 | ·64 | ·3072 | ·13107 2 | ·05242 9 | ·02013 | ·0075 | ·003 | ·82 | ·64 | 2·0172 | ·20547 2 | ·26061 | ·2244 | 2·128 | 1·99 |
| ·33 | ·66 | ·3267 | ·14374 8 | ·05929 6 | ·02348 | ·0090 | ·003 | ·83 | ·66 | ·0667 | ·28714 8 | ·37292 | ·3634 | 2·289 | 2·17 |
| ·34 | ·68 | ·3468 | ·15721 6 | ·06681 7 | ·02726 | ·0108 | ·004 | ·84 | ·68 | ·1168 | ·37001 6 | ·48936 | ·5093 | 2·459 | 2·36 |
| 0·35 | 0·70 | 0·3675 | 0·17150 0 | 0·07503 1 | 0·03151 | 0·0129 | 0·005 | 0·85 | 1·70 | 2·1675 | 2·45650 0 | 2·61003 | 2·6622 | 2·640 | 2·56 |
| ·36 | ·72 | ·3888 | ·18662 4 | ·08398 1 | ·03628 | ·0152 | ·006 | ·86 | ·72 | ·2188 | ·54422 4 | ·73504 | 2·8226 | 2·832 | 2·78 |
| ·37 | ·74 | ·4107 | ·20261 2 | ·09370 8 | ·04161 | ·0180 | ·008 | ·87 | ·74 | ·2707 | ·63401 2 | ·86449 | 2·9905 | 3·035 | 3·02 |
| ·38 | ·76 | ·4332 | ·21948 8 | ·10425 7 | ·04754 | ·0211 | ·009 | ·88 | ·76 | ·3232 | ·72588 8 | 2·99848 | 3·1664 | 3·251 | 3·27 |
| ·39 | ·78 | ·4563 | ·23727 6 | ·11567 2 | ·05413 | ·0246 | ·011 | ·89 | ·78 | ·3763 | ·81987 6 | 3·13711 | 3·3504 | 3·479 | 3·54 |
| 0·40 | 0·80 | 0·4800 | 0·25600 0 | 0·12800 0 | 0·06144 | 0·0287 | 0·013 | 0·90 | 1·80 | 2·4300 | 2·91600 0 | 3·28050 | 3·5429 | 3·720 | 3·83 |
| ·41 | ·82 | ·5043 | ·27568 4 | ·14128 8 | ·06951 | ·0333 | ·016 | ·91 | ·82 | ·4843 | 3·01428 4 | ·42875 | 3·7442 | 3·975 | 4·13 |
| ·42 | ·84 | ·5292 | ·29635 2 | ·15558 5 | ·07841 | ·0384 | ·018 | ·92 | ·84 | ·5392 | ·11475 2 | ·58196 | 3·9545 | 4·244 | 4·46 |
| ·43 | ·86 | ·5547 | ·31802 8 | ·17094 0 | ·08821 | ·0442 | ·022 | ·93 | ·86 | ·5947 | ·21742 8 | ·74026 | 4·1741 | 4·529 | 4·81 |
| ·44 | ·88 | ·5808 | ·34073 6 | ·18740 5 | ·09895 | ·0508 | ·026 | ·94 | ·88 | ·6508 | ·32233 6 | 3·99374 | 4·4034 | 4·829 | 5·19 |
| 0·45 | 0·90 | 0·6075 | 0·36450 0 | 0·20503 1 | 0·11072 | 0·0581 | 0·030 | 0·95 | 1·90 | 2·7075 | 3·42950 0 | 4·07253 | 4·6427 | 5·146 | 5·59 |
| ·46 | ·92 | ·6348 | ·38934 4 | ·22387 3 | ·12358 | ·0663 | ·035 | ·96 | ·92 | ·7648 | ·53894 4 | ·24673 | 4·8922 | 5·479 | 6·01 |
| ·47 | ·94 | ·6627 | ·41529 2 | ·24398 4 | ·13761 | ·0755 | ·041 | ·97 | ·94 | ·8227 | ·65069 2 | ·42646 | 5·1524 | 5·831 | 6·46 |
| ·48 | ·96 | ·6912 | ·44236 8 | ·26542 1 | ·15288 | ·0856 | ·047 | ·98 | ·96 | ·8812 | ·76476 8 | ·61184 | 5·4235 | 6·201 | 6·95 |
| ·49 | 0·98 | ·7203 | ·47059 6 | ·28824 0 | ·16949 | ·0969 | ·054 | ·99 | 1·98 | 2·9403 | 3·88119 6 | 4·80298 | 5·7059 | 6·590 | 7·46 |
| 0·50 | 1·00 | 0·7500 | 0·50000 0 | 0·31250 0 | 0·18750 | 0·1094 | 0·063 | 1·00 | 2·00 | 3·0000 | 4·00000 0 | 5·00000 | 6·0000 | 7·000 | 8·00 |

Formulae: $f(x+\theta h) = f(x) + \theta\tau + \theta^2\tau^2 + \ldots + \theta^n\tau^n + \ldots$

$f'(x+\theta h) = \tau + 2\theta\tau^2 + 3\theta^2\tau^3 + \ldots + n\theta^{n-1}\tau^n + \ldots$

where $\tau^n \equiv \tau^n f(x) = \dfrac{h^n}{n!}\dfrac{d^n}{dx^n} f(x)$, in which $h$ is the interval of tabulation in $x$.

(See B.A. Mathematical Tables, Part-Volume B, *The Airy Integral*, page B7)

Cambridge: University Press, 1946

For EU product safety concerns, contact us at Calle de José Abascal, 56–1°,
28003 Madrid, Spain or eugpsr@cambridge.org.

www.ingramcontent.com/pod-product-compliance
Ingram Content Group UK Ltd.
Pitfield, Milton Keynes, MK11 3LW, UK
UKHW050008151225
465965UK00013B/451